T0137043

The New Civilization Upon Data

Zipei Tu

The New Civilization Upon Data

How Big Data Reshapes Human Civilization, Business and the Personal World

Zipei Tu
Guangzhou, Guangdong, China

ISBN 978-981-19-3083-6   ISBN 978-981-19-3081-2   (eBook)
https://doi.org/10.1007/978-981-19-3081-2

Jointly published with China Translation & Publishing House
The print edition is not for sale in China (Mainland). Customers from China (Mainland) please order the
print book from: China Translation & Publishing House.

This Springer imprint is published by the registered company Springer Nature Singapore Pte Ltd.
The registered company address is: 152 Beach Road, #21-01/04 Gateway East, Singapore 189721,
Singapore

Preface: From Adolescence to Maturity and the Data Revolution in Our Era

A New Way to Divide Data

With the rise of big data, the power of data is often discussed, but the problem is that the data referred to is not always the same kind of data.

Nowadays, everything that is recorded is referred to as data, and that extends to written language. The logic underlying this is that data has become a concept. In the traditional sense of the term, data referred to the results gained by measuring physical things. What can be quantified can be turned into data. This could be called numerical data. However, "data" is now applied to images, video, and audio recordings—not things being quantified by measurements but recorded from the surrounding environment. This data is a type of evidence or record of events. What can be recorded can be turned into data. This could also be called recording data.

Numerical data and recording data were originally completely unrelated items, but they have been united by the fact that they can now both be stored as bits and bytes on a computer. When we talk about big data, I think the most precise way to sum it up is with this equation:

Big data = traditional numerical data (a quantifiable piece of data derived from measurement, like an exterior temperature of 28 °C) + modern recording data (produced by capturing the world around us, like a digital image).

Numerical data is closer to the original concept of data as a number, but, looking back over history, it was recording data that appeared first. Our earliest attempts to record history are an example of this latter type of data. Recording data is like a shadow cast by history. Numerical data came along later, produced by things like recording changes in the sky, the movement of stars, the location of mountains, or the course of rivers. This precision was gradually extended to many other things. The origin of science is trying to record data about physical things. Numerical data is the mother of science, you could say. It is based on precision.

Around the sixteenth century, when human beings embarked on a new age of exploration, there was a new wave of numerical data produced by the popularization of navigational instruments and the spread of similar inventions in the fields of land surveying, architecture, design, mining, and demographics developed. We discovered that quantitative data could provide solutions for science and management that were not possible with merely qualitative descriptions. The first wave of data was a wave of numerical data.

This was the dawn of humanity's first data spring.

In the twentieth century, there was an explosion of data following the spread of computers, the Internet, and smartphones. In terms of magnitude, this blew away what was seen in the sixteenth century. Judging by volume, the vast majority of big data is now recording data. It is data produced by the spread of technologies to record the world around us. This is what is meant by universal recording. If we look at Internet platforms, most of the data they host is recording data.

The previous 5,000 years of our civilization seem impressive compared to our comparatively brief modern age. But the present state of universal recording means that we know much more about it than we do in the past. Even if we have libraries full of history books, they tend to focus on the deeds of emperors and generals. Looking for more individualized records of people from behind that small coterie of great men, we find comparatively few. Universal recording changes that. In the future, there will not only be histories of states or societies or industries, but also an unprecedented number of personal histories. Historians of later ages will have an immense amount of material to study.

Of course, numerical data has things to tell us about Chinese history, as we saw with the discussion sparked by Ray Huang's ideas on "numerical management." I focused on numerical data in my previous two books, *Big Data* and *The Peaks of Data*.[1] The latter was intended as a continuation of Ray Huang's work. But this book focuses more on recording data.

An Unfinished Revolution

The Big Data Fever we see today is about a specific type of data: recording data. It has nothing to do with numerical data.

There's nothing strange about the interest in recording data. It can be explained by the changes that the Internet has brought to society and the fortunes that have been built from it. If we look at e-commerce, social networking, or media platforms, they have all benefited from recording data.

Almost all human behaviors can be recorded and converted into data. If the core of numerical data is accuracy, then the core of recording data is clarity.

The popularity of smartphones has made data ubiquitous, but most of us have only a superficial understanding of it. The discussion of the topic is mostly restricted to precision marketing, financial and social credit, and privacy.

---

[1] Editor's note: The simplified Chinese version of *The Peaks of Data* was released by CITIC Publishing Group in May of 2014.

Precision marketing refers to Internet advertising. Compared to traditional marketing, online platforms can now push personalized advertisements to users based on the data that they generate. Using consumer data for advertising greatly improves the efficiency of marketing. This has driven the commercial big data revolution. Whether we are talking about the Chinese BAT (Baidu, Alibaba, and Tencent) or Facebook, Amazon, and Google.

When it comes to data for credit purposes, the main users of the technology are financial institutions. This is another way that Internet companies can wring profits out of big data. It is also the reason why we have seen so many Internet companies entering the financial field. The business model is to use big data to assess customers through their purchase records; financial products are offered to customers based on what the company can learn from data analysis. Alibaba's Sesame Credit and Tencent WeBank's microloans are offered on the basis of credit scores.

Both precision marketing and credit scoring require monitoring customers and collecting data on their behavior. Individual users become the target of observation, analysis, and surveillance. This brings us to the third area of discussion: privacy. Now that Internet companies are collecting more data, we also hear more about that information being lost in privacy breaches. Tragedies have resulted from this.

Recently, a friend told me that he had received targeted advertising on Toutiao for swimsuits and goggles. They somehow knew that he had gone swimming. The only way that was possible, he decided, was that he had used his smart watch's "pool mode."

It seems clear this speculation was correct. Toutiao must have accessed the data from his watch. There is ambivalence about this sort of business model. On the one hand, we feel as if our rights have been violated, or at least disrespected. We might feel betrayed. Public opinion usually expresses dissatisfaction toward this business model in black-and-white terms and evaluates this kind of practice as unfair. But on the other hand, if he really needed goggles, it's quite an efficient system.

These two business models are feasible because they use recording data to understand consumer dynamics. The value of recording data to commercial and management spheres is the focus of this book. However, for the sake of convenience, I will not refer to "recording data" but simply "data."

This is big data as the public sees it. The first two concerns—precision marketing and credit—have popped up in public opinion, leading us to the third issue of privacy, but this is only the tip of the iceberg. Can the discussion really be confined to the rise of several large firms, improving efficiency, and rectifying the credit system? Is it so simple to get rid of the inconveniences and push for the positive use of technology?

I think we need to go deeper.

In the business world, big data is still subject to ongoing reforms; Internet applications are constantly updated and revamped. Big data is not being used to its full potential yet, but when it is, we will see a final form that allows complete automation of most commercial processes—that is, intelligent business. Commercial civilization will be completely recreated and reshaped by this process. That is what I hope to explain in this book.

The commercialization of recording data has led to the rise of big data, but business is only one angle from which to look at this problem. This big data revolution will be socialized. We will someday see smart business, but it relies on data, too. The great changes in social governance and individual lives will have an impact on all aspects of social life. This will push us toward an entirely new civilization and remake society. This is a story that might be told the same way in many parts of the world. We need to put aside the discussion of the details and zoom out to get a look at the direction things are going. The power of data will reshape society and maybe even human nature.

I call this Data Civilization. That is the theme of this book.

A New Civilization Rising

What is the meaning of "civilization"? Civilization is the gathered sediment of the flows of history. It includes the inventions and innovations that have stood the test of public opinion over time. It gathers together these essential elements: spoken and written languages, tools and instruments, morality and ethics, beliefs and religious practices, laws, family systems, the *polis*, and the nation state.

Our present written language is also a type of data, or we could say that data also includes but surpasses written language. The written word is only a subset of data. We could make this analogy: the written word is to data as gold is to metal.

Language once shaped human civilization but data has taken its place. It has the power to shape psychology, ethics, and even religious beliefs. I hope this book will explain that.

The Chinese word for civilization is *wenming*, which combines the characters for "language" or "culture" with the character for "understanding" or "light." Humanity has always been grasping for that "light." Confucius said: "If I could know how to realize my ideal in the morning, I would gladly accept my death in the evening." In other words, if someone arrives at a revelation about the truth, they can die happily. Death is a price one should be willing to pay for the truth. In Chinese, all of these words contain the "light" of civilization: wisdom—*mingzhi*—virtue—*mingde*—reason—*mingli*—and when we say a person is sensible, they are *mingbai*. Finding this "light" is the quest of humankind. The "light" of civilization illuminates all things. It is the prerequisite for all material existence and development. When we rise above the primitive state of subsistence, we call this civilization. As the Chinese word—*wenming*—suggests, it is "language" that brings us to the "light."

Now, we can go beyond "language" as a source of this "light." Data casts its own illumination. Some things that were historically hidden or obscured have become apparent because of the light of data. New knowledge has become clear. This allows us to put it to use. Humanity has never understood or managed itself with as much definition, clarity, precision, or objectivity as at present. This is because of data. Data allows us to uncover many facts that were unknown. The idea of trusting things to luck in an uncertain world will come to an end. We are entering a more civilized and safe era.

There is a story in the Bible that says people in every part of the world used to speak the same language. It was easy for everyone to communicate and cooperate. That led the people of the Earth to pool their talents together and attempt to build a tower to heaven. Humankind united for the project and the tower grew higher and higher, until God grew anxious. He saw that their project was only possible because they spoke the same language.

Construction of the tower was halted when God suddenly gave each group of people their own language. From that moment on, mankind was divided and unable to cooperate effectively. The Tower of Babel failed. The people scattered.

Whether we are talking about numerical data or recording data, it is all unified under the 1s and 0s of binary computer language. According to the logic of the Bible, we could unite to build our tower to Heaven.

There will be even greater changes on the social and national levels. More individual behavior is being recorded, which gives states the power to track citizens, vehicles, and other things. If a well-developed bureaucracy can master the use of data, it will have an increasingly defined image of the societies they govern. Data creates a high definition society, and this book argues that effective governance can only come through being able to see with clarity.

My work builds on Ray Huang's ideas about "numerical management." I believe that China's relative backwardness over the past century was the result of many people forsaking precision. There was a lack of numerical data in Chinese history.

When compared to numerical data, recording data gives the state even more tools and methods for governance. The best way to modernize governance is by using big data. I believe, when we're talking about China, that the *only* route to the modernization of governance is big data.

In Chinese history, there was always a debate about the rule of virtue versus rule of law. Rule of virtue is based on love for one's fellow man, while rule of law advocates for a strict legal code. This debate has been raging for the past thousand years. It's clear that China now wants to establish a society based on rule of law. But this book proposes something in addition to that: rule of data, or data governance. Rule of data means the efficient collection, use, and analysis of data to govern the country. It is also referring to the way in which data itself can be governed. Effective management of data for the effective management of the state will be an important component of governance.

Big data in the context of the Internet represents a new force in national affairs and it is changing traditional society. We can look at the example of the spread of surveillance cameras, which can capture and replay events that would previously have gone unnoticed. The idea of trusting things to luck has its limits now. The crime rate will go down. Data governance addresses a fundamental human need for security.

But data and the Internet bring new challenges, as well. It is difficult for state power to be projected into every dark corner of the Internet. Now, more functions of the state are being transferred to private companies and civil society organizations.

While you would have once found the information in a more roundabout way, the reputation of a business can now be found by reading customer reviews. If you see a news report about a disaster and want to make a donation, you will probably go to an online platform instead of directly to a traditional charity. A government department can launch a website for its services, but the real revolution in public service has come from integrating services into commercial Internet platforms. The media provides more examples. Toutiao has supplanted TV stations and newspapers as a source of information. But that means that private firms can now challenge the government in the realm of public opinion. How do we restore the credibility of state power? It seems that integration with private Internet companies is the only solution.

How do we use the power of big data and the Internet and also restrain it when necessary? Data governance will give birth to a new political civilization. That is the challenge we face. If data governance is done well, China will have a great advantage in a new era. That holds true for the central government, as well as subnational governments and companies.

A Reliable Road to Success

Data civilization has meaning not only for states and societies but also for the individual.

Individuals need to grasp where the future is headed and what that will mean. They need to know what sorts of businesses will disappear and be born in their lifetimes. We can't allow people to stumble forward in the dark. If we want people to rush into the future and appreciate the scenery along the way, they need a light.

I believe that data civilization will mean that the potential of individuals will increase. The type of person that will succeed in a data civilization will have awareness and understanding of data and the talent to make use of it. A new era of data civilization will shape the destiny of individuals. We should adjust our ways of thinking.

When I was a kid, I remember hearing stories that would usually begin something like this, "A long time ago, far away, on the other side of the mountains..." I was fascinated by that description. This is a basic storytelling technique that creates a narrative space by distancing the person hearing the story from the events. The storyteller is describing a place that the person hearing the story has never been to, has never seen, has never experienced, and doesn't understand. So, the storyteller has the authority. In ancient times, it was old people and people that had traveled widely that were considered wise. They had seen and experienced times and places that most people knew nothing about.

The reason these places are unknown is because there are no records of them.

That has changed. There is no longer an unknown place on the other side of the mountains. Google Maps and Street View can take you there without having to leave home. People are recording their world in all places and at all times. There is no limitation set by time and space.

The Tang Dynasty poet Yuan Zhen was a gifted scholar. He wrote some lines of love poetry that have been passed down for generations: "All water is forgettable when you've seen the vast blue sea/No clouds so wondrous as those at Mount Wushan." He was writing those lines to mourn his wife. But the real Yuan Zhen was not quite as devoted as his poetic alter ego. Before marrying his wife, he had a lengthy affair with his student, which was detailed in *Yingying's Biography*, widely considered as the forerunner to later Ming and Qing Dynasty works, like *A Dream of Red Mansions*.

Yuan Zhen traveled widely and always had affairs. He is reported to have broken the hearts of Liu Caichun, a poet from the area around Shanghai, and also Xue Tao. The letters that Yuan Zhen and Xue Tao passed back and forth are beloved by literary scholars. Xue Tao was hopelessly devoted to Yuan Zhen, but he did not share her commitment. The gulf between the poetic character of Yuan Zhen and his true identity was very large.

But nobody is perfect. The same holds true for all the kings and generals recorded in history. But the lack of records means that we don't know the truth. We only have the beautiful traces left behind. Later generations have turned their ancestors into legends and demigods. Guan Yu became a god of war mostly because of myth. Wang Xizhi became the most famous calligrapher in history despite the fact that there were no surviving examples of his work to examine. The creation of gods and sages out of historical figures is all down to the lack of authentic records. The means and methods to record the truth simply didn't exist. The situation is now completely different. As long as someone has a phone, they can take pictures and videos, jot notes, and share what they record. That is what I mean when I refer to universal recording.

In an era of universal recording, it is hard to deify a person, let alone a figure as flawed as Yuan Zhen. Anyone that is subject to universal recording might show themselves to be a true hero, but their dark side will also be exposed. We now realize that even great sages are not much different from regular people.

Everyone now knows the rocky start that Jack Ma's career got off to, but if there was no record of this time, it would be hard to imagine. In earlier times, he may have been deified as a god of business. But with our modern recording tools, we get the truth. Recording data can bring great men to the level of ordinary people and erase the space between us and them. This is not historical nihilism but authentic history. The Chinese have a tradition of ancestor worship and that element in our culture has ensured the continuity of Chinese civilization, which should be taken as a positive feature, but data civilization allows us to see another level: legendary figures were the same as us. So, there's no need to have an inferiority complex. Even if we might not seem particularly outstanding to others, we can still go on to great success.

Having access to records can cause disenchantment. But the meaning of the word "disenchantment" is not entirely negative: it means breaking the spell—the spell that mystery and the unknown hold over us. Once we know that our heroes are mere mortals, it might give us mere mortals courage to be heroes.

But if the argument is that ordinary people can all be successful, what does it really take in an age of data?

This book attempts to find the "golden thread" in social and commercial civilization. I believe this "golden thread" can also be taken advantage of by individuals. It is a good way to find professional and career success.

I recall later in this book the time when I started my work life. In 1996, I joined the People's Armed Police and was humiliated by my commanding officer for my lack of ability to write reports. I was always more into science and technology, so it was hard to master the new skill, but I finally did it through a very simple method: I began clipping materials, filing them, summarizing them, and reading them to myself. Very quickly, my writing skills improved.

I believe this experience proves that the differences between people are far less than most imagine. Everyone is mortal. Everyone has hopes and dreams. Everyone is foolish in their youth. To accomplish things requires hard work. A famous line goes: "Is there anyone who was born to be a king or a general?" You can repeat that with confidence.

Recording is one path to individual success. When it comes to recording events, our brains are inferior to computers. Human memory is a differential mechanism and computer storage is an integral mechanism. Study requires that integral mechanism. We need to make good use of recording and data to improve our chances of success.

Whether we are talking about individuals or nations, the route to success sometimes runs parallel. One place this is true is the process of recording. Universal recording is the driving force behind personal success and also civilizational change. That civilizational change will in turn benefit the individual. I hold that data is going to change every part of our civilization. There are examples in the way that Alipay has changed traditional banking and WeChat has changed communications. Data civilization will create new rules based on superior models. There will be new languages, new paradigms of public order, new customs and cultures, and maybe even new core beliefs. To our society, the transformative power of data is like the introduction of agriculture or the steam engine to our ancestors. Data is undeniably giving birth to a new concept of civilization.

A New Concept of Privacy

It's undeniable that a new concept of civilization will also create some problems.

When we look below the surface of big data, we see that it is being used for private profit and social control. Internet giants are gobbling up great masses of data and reaping profits. These companies have grown to mammoth proportions in just over a decade. These companies were originally beholden to their masters but have now shown themselves to possess their own power. In China, as well as the United States and Europe, personal data has been collected without permission and used without authorization. In March of 2018, Facebook suddenly revealed that it had improperly bought and sold this data, perhaps allowing it to be used as a tool for political manipulation. The data that consumers create is turned against them by the Internet companies that collect them. Many of us find our personal lives enclosed in a physically small space where we are kept entertained by personalized news feeds and individualized shopping recommendations. These platforms provide comfort, but the data we provide to them might be becoming digital shackles.

These are two paradoxes of our new civilization. On one hand, the better and more comprehensive the data, the better and more comprehensively we can be marketed to, so we can have precision advertising, individualized home pages, and personalized services, but that can also lead us to be trapped within a narrow virtual space, fed a thin stream of information. On one hand, big data connects more people, but it can also compromise our privacy and rights. The Internet is increasingly monopolized and the individual has become a lonely skiff on the open ocean. We think we are the captain of our ship, but we are actually trapped. We can enjoy the sunrise, but each day is about the same as the last.

How should we address the problems of our new civilization? It is easy to become pessimistic or disappointed. I think there is a danger in letting these concerns become overblown. We run the risk of not being able to see the forest for the trees.

Conflicts always arise as civilizations develop. No civilization arrived in a perfect state. In China, we saw the contention of a hundred schools of thought, centuries of war, and forced integration during the Spring and Autumn and Warring States period. That was what was required to create a unified civilization in the Qin and Han Dynasties and establish Confucianism. Civilization is never born in its ultimate form but must evolve into that state. Civilization has to be accepted by the majority and become a kind of identity and belief system. The development of advanced technologies provides new possibilities for civilization, but the final form has not yet arrived.

The paradoxes discussed above might be resolved quickly. As cryptocurrency is popularized, blockchain technology is improving. In the future, an individual's data won't be stored by Internet companies, like Facebook, Alibaba, and Tencent, but on the blockchain, so they will need to obtain consent and authorization for the access and use of personal data. All we can say for sure is that the data revolution is still ongoing.

Even the privacy issue might be solved by the emergence of solutions from artificial intelligence. But we also need to develop a new concept of privacy.

For example, the issue of individualized pricing that is criticized in public opinion is made possible by algorithms that don't actually publicize any of our personal information. Toutiao, for example, infers our preferences based on our browsing history and clicks, then tries to make guesses on what we would like to see. It may seem as if our privacy has been compromised but no human eyes ever actually need to see our private data, since it is all accomplished by algorithms and AI. These mechanisms operate automatically and provide rapid matching with minimal human intervention. Should we consider our privacy to have been compromised by these processes?

This is the way things work now: data is processed by algorithms and AI. It is not being seen by anyone at all. Privacy is not violated by human intervention. There will be cases of people leaking personal data, but they will be rare. That suggests a new question: Does our personal data need to be kept secret from algorithms?

We don't mind being observed by the natural world, so why should we mind algorithms watching us? Maybe we shouldn't. In the future, algorithms and AI will be as natural as anything in our living environment. The use of our data is likely the only option to usher in a new age of AI, robotics, and human-machine hybrid intelligence.

Our new concept of privacy will draw a boundary around what we allow when it comes to the processing of our data by algorithms.

I believe the continuation and spread of digital civilization must be seen on a time-line of centuries or millennia. We need to approach data civilization with a century plan or millennium plan for humankind. Whether we are talking about states or individuals, we need to keep up to date with what is happening with the development of data civilization. This book is my attempt to look at how data and AI can develop human civilization and provide some guidelines to examine how it might develop.

Guangzhou, China   Zipei Tu

Preface

Data has existed since ancient times, but, now, with the emergence and popularization of the Internet, as well as the power of computers to store and process digital information, we are dealing with what is known as big data. In recent years, it has become part of everyday life, production and management, as well as social governance. It now stands alongside capital, labor, and natural resources as one of the key elements of human society.

The emergence of this new element stands to have an impact on politics, economics, and culture that is equal to the discovery of the New World in the 15th century. It deserves serious analysis and attention from people in all walks of life.

Although China got a slower start than the United States and other advanced industrial states, its large population, the size of its economy, and the rate of its development has given it an edge in development and application of big data technology. We have already seen the emergence of Baidu, Alibaba, Tencent, and Jinri Toutiao as leaders in the field, accompanied by many other tech unicorns. The Communist Party of China and the government of the country have also proposed a "National Big Data Strategy," which follows on the heels of the "Action Outline for Promoting the Development of Big Data," the "Big Data Industry Development Plan, 2016–2020," and other guidance documents. The development and application of big data technology is in a state of ascendancy.

Tu Zipei has emerged as China's first authoritative voice on big data and *The New Civilization Upon Data* is his third masterwork on the subject, analyzing the opportunities and challenges of this crucial technology to all aspects of human life, society, and civilization. He glides through history and literature, explaining the profound in simple terms. He incorporates a vast quantity of information in his expansive analysis. There is a lot to get out of the book, and, once you begin, you will find it hard to put down.

Tu Zipei does not merely prophesies the arrival of a data economy in a way that is unique and farsighted. There is little doubt that big data will become part of the global economy and its role in the Chinese economy is already clear, but Tu Zipei draws our attention to the shadow that this may cast over the economy and society. I hope that this book is the beginning of a trend of more research into and discussion of limiting the negative impacts of big data and magnifying its positives.

Justin Yifu Lin
Director
Institute of New Structural Economics
at Peking University

Chief Economist
World Bank

Contents

Chapter 1
Data Equality: The Impact and Original Sin of the New Business Culture

Abstract Civilization forms when rules are set to govern it. The source of our digital civilization's transformative power is business. This chapter begins with the achievements and setbacks of the new business culture that has propelled digital culture, and it contains a basic argument about how data rights should be treated. Just as with land, there's a simple question about this new asset: should it be privately held or publicly owned? The answers to that question could serve as an ideological or doctrinal litmus test; those answers represent choices for the path that a digital civilization might take. If data can be a privately held asset, existing business and finance structures will be transformed, along with our own personal lives.

1.1 As a Civilization Reaches Its Peak, the Fears of Its Collapse Grow More Terrifying

In March of 2018, the world woke up to shocking news about Facebook, the world's largest social networking site. It turned out that a small, obscure firm called Cambridge Analytica had gained access to the accounts of more than 87 million Facebook users.[1]

The flood of vicious criticism that Facebook endured is all the more understandable if you know the role it plays in the everyday lives of Americans. For the Chinese, it would be equivalent to Tencent's WeChat or ByteDance's Toutiao.

Facebook is not simply a tool for connecting with friends but its interface is also a place to receive information, news, and advertising. It would not be inaccurate to call Facebook a hybrid of WeChat and Toutiao. As news has broken about lapses in the protection of data at WeChat and Toutiao, the problems at these firms looked like mirror images of what Facebook had already been through.

In the wake of the controversy, Facebook agreed to comply with the European Union's General Data Protection Regulation (GDPR). Among other stipulations about data collection transparency, it also meant that users would be given far more

---

[1] At first, the number of users affected was believed to be 50 million. But on April 4th of 2018, Facebook CTO Mike Schroepfer announced the much higher 87 million number.

control over how their data was used, and it required companies using the Facebook platform to ask the consent of users receiving their content.

The 2018 Facebook controversy serves as a prelude to what I believe will be major changes. The core issue is no longer privacy, but data rights. This is something that is central to the business model of most Internet companies. It goes beyond a legal issue to become a question of public welfare. At the moment, most people have not yet realized that their data rights are being violated.

In light of the data rights issue, Internet companies must now rethink their business models. If the public at large claims the right to their own data, the Internet will embark on a new age. That new age will see a massive restructuring of the business landscape. It will completely change the views that people and governments previously held about the regulation of Internet companies.

It's undeniable that the earlier age of the Internet created a new paradigm in finance and business. We have watched the rise of a new economy. There have been many terms batted around, like "information economy," "internet economy," "digital economy," "smart economy," but I think the more accurate term would be "data economy." Vast waves of data are crashing down on business and politics. More and more daily decisions are being turned over to data and algorithms. Although they might provide some relief, these waves still carry a sting. As civilizations develop and change, humankind is destined to meet with victories and setbacks, but also moments that are seemingly inexplicable. The most unshakeable fear is the fear of the unknown.

1.2 Psychological Invasion: The Birth of Big Data "Mindreading"

The extensive use of big data (a term for vast sums of data but also the systems that control it) in election campaigns can be traced back to Barack Obama's campaign to be the 44th President of the United States. In 2008, during his first run for President, he established a personal campaign website (BarackObama.com) that allowed him to collect the email addresses and personal information of thirteen million individual voters. The Obama campaign decided to steer clear of the tried and true marketing and publicity strategy of "mass mailing"—blanketing voters or consumers with the same message. Instead, the campaign took the information gleaned from the website and fed it to a team of data scientists that used it to classify voters and tailor a message specifically for their group. This method helped him to win the presidency in 2008 and in 2012 he was up for re-election. In the preceding four years, Facebook had reached 800 million users and the campaign made use of that audience by teaming up with the site. When a voter logged on to the personal website of Barack Obama, they were prompted to enter their Facebook account details, give permission to the Obama team to read information from their profile, as well as giving them the authorization to make posts on their behalf.

Both 2008 and 2012 elections were large Democratic Party victories, and they became textbook cases of how harnessing large amounts of data could win elections. Republican candidates were envious of the technological advantage the Democrats had in mobilizing voters. I once summed it up like this: "success in politics now comes down to technology."[2]

After that, the Republicans learned from their painful lessons. Individual Republicans began to reach out to people in the tech sector and invest in their companies. One of those firms was Cambridge Analytica, which had figured directly in the 2018 Facebook controversy. Donald Trump advisor Stephen Bannon had served on the board of the company, and Republican money man Robert Mercer had invested heavily.

The Cambridge Analytica name came in part from the involvement of Aleksandr Kogan, an American of Moldovan ancestry that had worked as a research associate at the University of Cambridge's Department of Psychology. In 2015, Kogan launched a Facebook application, called "This Is Your Digital Life." The app was a simple personality test that also came with the offer of an approximately five dollar prize. This is another interesting thing about our time: few would stop to pick a quarter up off the sidewalk, but people will jump at the opportunity for a minimal cash prize online. In this case, it was enough to attract more than 320,000 users to participate. It's time we demystify the idea that anything on the Internet is "free." If you aren't paying actual money for it, you're giving up something even more valuable to Internet services—you're giving them your data.

That was what Kogan was up to. His app collected the data from the personality test, but it went on to scrounge for a wide range of personal information stored by users on the platform. That included information about location, Facebook friends, and what content they had liked on Facebook.

The data was then downloaded to a server owned by Cambridge Analytica.

But how did Cambridge Analytica go from 320,000 users' data to 87 million? That's an increase of more than 200 times.

The reason for this is that the default privacy policy setting on Facebook. If users did not specifically change those settings, any status changes or posts to their wall would be visible to Facebook friends. For Kogan's app, that meant that he would have access to everything posted by the friends of those 320,000 original users. The fact that most users had not changed their default permissions meant that Cambridge Analytica could suck on the data of anyone connected to the first group, even if they did not access the app. The cost of running this Facebook data collection scheme was a mere $1.6 million.

That begs the question of how much Facebook knew about Kogan's app.

It appears they were fully aware. It only took Cambridge Analytica a matter of weeks to harvest the data of 87 million users. Had Cambridge Analytica not had prior approval, the rate at which they were algorithmically siphoning data would have set off alarm bells. Facebook does not normally allow app developers to harvest that

---

[2] Tu (2015).

amount of data, but Kogan had received their blessing with the claim that it was for academic research, which they have always expressly allowed.

Eventually, though, Kogan transferred the data to Cambridge Analytica cloud servers, which meant he had violated his agreement with Facebook. Whether or not Kogan's claim of academic research was a ruse or not, the problem was that Facebook had no way of tracking where data harvested by researchers or developers ended up. If not for the controversy that erupted over Facebook and Cambridge Analytica, there's no telling what would have happened. The data could have been bought and sold countless times. The fact that data is currently untraceable is one of the key problems with our new paradigm, and we will need to come up with a solution.

Facebook was also particularly careless with the data that Kogan was harvesting because none of it was considered "sensitive." Kogan wasn't grabbing passwords or other information that would allow direct access to a user's account. He was harvesting seemingly insignificant information about user locations and preferences. That information was generally publicly available on the platform.

Cambridge Analytica took everything that Kogan had harvested and began using algorithms to identify patterns. They were not trying to classify voters based on traditional demographics, but based on their psychology. This new big data method paved the way for Obama's two election victories.

Digression: Psychological Analysis Through Data

Psychologists have always believed that human traits can be described in human language. It holds true for all languages that the more important a characteristic is, the more vocabulary will be devoted to describing it. This is called the lexical hypothesis—the idea that more socially important traits will show greater density in the lexicon of a language. This is an idea that goes back to the nineteenth century, but some of the most important work in the field goes back to American psychologist Gordon W. Allport (1897–1967). He identified 17,953 words describing personality and behavioral traits. He further classified these words according to their type. His categories line up quite well with the taxonomy of the Big Five personality traits (often abbreviated as OCEAN), which were suggested in the 1980s.

The Big Five divides traits into five categories:

- Openness to experience is characterized by inventiveness and imagination;
- Conscientiousness is characterized by self-discipline and the pursuit of perfection;
- Extraversion is characterized by a tendency to be outgoing;
- Agreeableness is characterized by a cooperative spirit;
- Neuroticism is characterized by sensitivity and anxiety.

The combination of big data and psychology was immensely powerful. Researchers have found that a person's psychological traits can be reliably determined by analyzing data from their social media profiles—and this method can be even more accurate than analyzing statements about the subject from their family and friends.

With enough data, algorithms can get a very good idea of a user's psychological traits. Even with limitations, like only having access to data on what the user has "liked" on a social media platform, algorithms can begin to create an accurate profile. Every "like" is given for a reason, after all. After analyzing ten Facebook "likes," algorithms know the user better than their acquaintances know them; after seventy likes, the algorithm has a deeper understanding of the user than even their friends; with access to 150 "likes," the algorithm knows the user better than close family; and with 300 likes, the algorithm knows the user better than their spouse.[3]

Cambridge Analytica went even further. They took the data of 87 million users and compared it to available data on the consumption habits of 220 million Americans. Combining that with their Facebook data and building on it, they could determine a potential voter's gender, age, hobbies, psychological characteristics, occupation, and political inclination. They began dividing these potential voters into hundreds of categories, determining what kind of messaging would best play on their hopes and fears, and figuring out what would generate a sympathetic response and what would trigger their "inner demons."

This was the birth of big data "mindreading."

If you discover a person's "inner demons," you can then calibrate the information being delivered to them. You can package content that scratches an itch they might not even know they have. Completely unobtrusively, you can change someone's choices and judgements.

Thinking back to the election of Donald Trump in 2016, we can see this in action: the hot-blooded patriots were fed content calling on them to help "Make America Great Again," and more educated, wealthier users were fed content that sold Trump through a more rational, intellectual argument. Algorithms could be used to identify social media users that were inclined to Trump but unlikely to vote or donate money, and then they could be pitched a call to action depicting the outcome of Hillary Clinton taking power. To people on the fringes, the content pushed to them would be heavy on conspiracy theory and hairraising clickbait articles. At the same time, Democratic voters could be targeted with negative content about the Clinton family, potentially demoralizing them, increasing the likelihood that they stay home on election day.

The content pushed to potential voters would be carefully tailored for their target subgroup. Of course, some of it would be "fake news." This content was designed to trigger something buried deep in human nature, according to what the algorithm knew about the psychology of the user. At the same time, the way social media platforms were structured, there was a low likelihood that an algorithm would push any contradictory information, creating an echo chamber of sorts: the user would have no awareness that legitimate dissenting opinions existed.

---

[3] Youyou et al. (2015).

The content was formulated like a pill, and administered in much the same way: potential voters would receive the same messages many times, until they finally took effect, guiding them to give the desired response.

Essentially, this is applying Internet marketing tools to elections.

Modern advertising interest has a history stretching back a century and it's always been intertwined with the development of psychology. From the early twentieth century, advertisers used statistical analysis to figure out how to get people to part with their money. Like Thomas Edison testing 1600 potential lightbulb filaments before discovering tungsten, advertisers test the efficacy of new methods through trial and error. Psychology shows us that there are many "loopholes" in the superficial rationality of human psychology. People can be influenced by emotion; it is quite elementary to exploit this tendency. After a century of research, the advertising industry has discovered exactly what weaknesses hide in human nature. In his study of emotional manipulation in advertising, Nobel Prize-winning economist George A. Akerlof called this "phishing for phools."

The Internet has greatly increased the efficiency of traditional advertising in several ways.

First, it has greatly spread up the A/B testing required to find the best lure when "phishing for phools."

An e-commerce platform can serve multiple versions of the platform to different customers, then analyze the result. For example, say that they have a million users, they can serve version A to 500,000 users and version B to 500,000 users. When it comes back that version A saw 10,000 purchases and version B saw 120,000 purchases, it's clear which is delivering the higher conviction rate. After that, version B can be pushed to all users.

This is a simplification, of course, since there will be more than two discrete versions. Smaller changes can be made, like to headline text. A site can check whether a question mark or a full stop generates more clicks on a headline, for example. The font can be changed, too, made larger, or the color can be changed. The layout of the site can be changed, too, with headlines, banners, and columns being moved around. The number of possible iterations is nearly infinite. The process of AB testing can be repeated until the site is completely optimized. At that point, "phishing for phools" approaches perfection. When the results of the testing are served to customers, the click-through and conversion rate increases.

Artificial intelligence can adjust, track, and compare changes in a way that humans cannot. AI algorithms can carry out and quantify this testing, while also simultaneously adjusting and pushing out fresh changes. This work is served up to countless millions of shoppers, usually without them perceiving it.

Second, the Internet allows for different content to be served to different users.

The greatest shortcoming of traditional advertising is that one identical piece of content is transmitted in roughly identical form to an audience. No matter how deep the advertiser's understanding of human nature was, they had to work with general tendencies. No single advertisement will appeal to everyone in the audience. Advertisers are "phishing for phools," in a big lake; not every "phool" responds to the

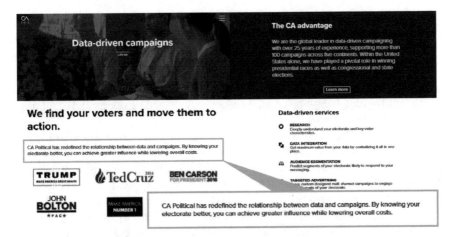

Fig. 1.1 The Cambridge Analytica website: "Data-driven campaigns"

same lure. The Internet allows advertisers to serve different bait to "phools," exactly matching their tastes.

The Internet allows things that traditional advertising could not. The Internet can track customers' every move, analyzing trends in their browsing or changes in their hobbies. Once it knows the customer, it can rapidly change the content that's being pushed to them. Even when serving the same content, it can make subtle changes to better appeal to the traits of the marketing subgroup. The algorithm can add a coupon for exactly the product it knows the customer might purchase, or it could finetune the time or frequency of notifications. This is what's called a "precision push."

Researching the likes and dislikes of people is a science. To finetune content and put it in front of users at precisely the right moment is a matter of technology. Big data is where science and technology meet.

That was what Trump was looking for when he hired Cambridge Analytica.

The Trump team contracted Cambridge Analytica five times. We can see from election finance disclosures that it cost them a total of $5.9 million.[4] What "data pill" did Cambridge Analytica mix up to help the populist campaign of Donald Trump defeat Clinton's Democrats? What role did Cambridge Analytica have in swaying the result? There's no way to know for sure. The psychological impact of the content that Cambridge Analytica may have created for the campaign is hard to prove or quantify.

On the homepage of Cambridge Analytica, they claim to have taken part in more than a hundred elections around the world. That list includes Brexit in the United Kingdom. This turned out to be an unprecedented black swan event. The world woke up to a narrow—52 to 48—victory for Leave voters. The effect of "data pills" might be hard to quantify, but we can be sure they have an outsized effect on close elections (Fig. 1.1).

---

[4] "Summary data for Donald Trump, 2016 cycle," opensecrets.org.

We find your voters and move them to act.

We are the global leader in data-driven campaigning with over 25 years of experience supporting more than 100 campaigns across five continents. Within the United States alone, we have played a pivotal role in winning presidential races as well as congressional and state elections.

As I said before, I don't believe the controversy around Facebook and Cambridge Analytica is about privacy.

In Facebook's final statement on the matter, the company did not apologize for data theft, but rather a failure to protect users' data. This was not a matter of "leaking" data, either, since users had consented to the firm collecting the information, and much of what Cambridge Analytica harvested was already public. It is not unfair to call it an invasion of privacy, but the real problem is far more serious.

The Facebook-Cambridge Analytica scandal showed that by harnessing publicly available data, the Internet can become a tool to influence, control, and dominate psychology and ideology. This is not merely an invasion of privacy, but an intrusion into the hearts and minds of people around the world.

The controversy was even more embarrassing and unfortunate than it might have appeared on the surface, since Facebook was aware that Kogan was misusing data. In 2015, they had suspended his account and asked Cambridge Analytica to delete information they had harvested. The order, however, was not enforced in any way. Kogan merely checked a box on an online form and sent it back to Facebook.

None of the data was deleted. One unique thing about data as a resource is that no company or individual can effectively assure its destruction. Tangible items can be given and then taken back—or there can at least be proof of the items' disposal. But data is nothing like that. Once data is provided to someone, it can never be recovered.

The Facebook controversy was centered on the United States, but there were similar events in China in 2018. These were not privacy issues, but nor were they exactly similar to the manipulation practiced by Cambridge Analytica. These controversies with Chinese tech firms had a direct impact on users' pocket books.

1.3 Price Manipulation: Big Data Price Discrimination and Algorithmic Schemes

In December of 2017, a user of microblog service Weibo recounted their experience of price gouging on a travel website. He had regularly booked the same room and knew that the price fluctuated between approximately 380 and 400 Chinese yuan. One day, he was about to make a booking at a rate of 380 yuan a night, but just by chance, decided to inquire directly with the hotel. The front desk informed him that they were charging an off-season rate of 300 yuan. He logged back onto the travel website with a friend's account and saw that it was displaying the 300 yuan rate. On his own account, it was still 380 yuan.

廖师傅廖师傅 ✿

2017-12-29 07:54 来自 iPhone 7 Plus

最近发现很多"聪明"的互联网企业利用大数据"杀熟"已经成为了一种常态。本人两
次亲身经历跟大家分享一下：

一、经常通过某旅行网站订一个出差常住的酒店，长年价格在380-400元左
右；前几天在该旅行网站用差不多的价格又住这个酒店，办入住时好奇的问了一
下前台现在酒店的价格，她说现在是旅游的淡季，价格很低，差不多300元左右。
我让朋友用他自己的账号查了一下，果然是300，然后我自己用本人的账号去查，
还是380。我打电话问客服人员，接线员告诉我，可能是我缓存的问题......我忍住
没骂人，告诉他如果不解决这个问题，我会起诉他们欺诈。最后他们用最快的速
度免了这间房的房费。

Fig. 1.2 Screenshot of the account of Weibo user Master Liao Master Liao[5]

This touched off a flurry of similar complaints on Weibo. "I took a taxi to meet up with some classmates," one user wrote, "and, despite the fact that our route and vehicle were nearly identical, I was charged five or six yuan more."

Another Weibo user wrote: "I booked a flight online but didn't confirm it, then ended up canceling. When I went back to make the booking again, the price had nearly doubled."

When *Science and Technology Daily* reported on the situation, they called it "big data gouging."[6] The Chinese term they used might literally be translated as "big data kills the people they know best"—this is hinting at the knowledge that e-commerce sites have about their users, the result of harvesting their data (Fig. 1.2).

A few days later, *Shanghai Morning Post* conducted their own tests, proving that several travel and e-commerce platforms were using the same methods. They found that platforms could tell when shoppers were in a hurry—and then offer them a higher price.[7] Anhui TV's *Social Transparency* ran a censurious report on March 24th, 2018:

There's a saying about merchants sizing up shoppers before they give their price, and now there are examples of it happening online. For example, ride-hailing services were giving different prices to different customers for an identical route. For one, the price was 18 yuan, and for another user, it was 11 yuan. These users noticed the seven yuan price difference because they worked for the same company, lived in the same off-site dormitory, and often used the ride-sharing service to get to work. They discovered that there was a routine seven or eight yuan difference in the price.

What's going on here? It turns out that they were using different phones. Those using iPhones routinely got prices 30% higher than those using Android phones. There's been analysis done of streaming platforms that show the same pattern of overcharging iPhone

---

[5] Image source: Sina Weibo.

[6] Zhai (2018).

[7] Song (2018).

users. On one site in particular, a user logged in with their Android phone and got a subscription fee of 178 yuan a year. When they logged in again with an iPhone, the fee listed had increased to 218 yuan.

Basically, what the commentary is saying is: the platform could determine what phone a user was accessing the site on, then list higher prices for those using high-end devices. This is not unlike a shopkeeper gouging the guy that they see stepping out of a BMW in a mink coat. The line about "big data kills the people they know best" could be changed to "big data kills the rich." The logic here is that the wealthier should be charged more. If Jack Ma comes in to buy an item, you should charge him more than you would his driver.

According to preliminary data from the China E-Commerce Research Center, this is a widespread practice, used by major platforms, including DiDi, Ctrip, Fliggy, JD, Meituan, and Taobao Dianying. The problem is particularly severe on online travel sites. When confronted, the reaction was not uniform, with some sites outright denying the phenomenon, others refusing to comment, and a few admitting that there was some use of algorithms.[8]

From my understanding, the idea of using big data to adjust prices goes back about five years. This recalls what I said about advertisers suddenly having the ability to push a thousand variations of an advertisement to a thousand customers, rather than pushing one version to a thousand customers. By 2013, it was becoming clear that more and more shopping would be done online, and large platforms like Alibaba's Taobao were experimenting with the idea that each customer would be served a unique page. For this to work, the page needs to be served quickly but accurately. Finding that balance is what Alibaba has been working on. The experience of opening the platform needs to be as seamless as a customer walking into a brick-and-mortar location and seeing the items he wants stacked up beside the entrance. This is something that is only possible in the realm of e-commerce (Fig. 1.3).

I had three of my colleagues open Taobao on their phones at the same time. It's easy to see the differences in what the platform served them.

Left: The search box suggests "Canon 6D DSLR." The banner below it shows a pre-cooked chicken with the slogan of "Ingenuity & Flavor." Below that, there are recommendations listed under "Selections from Hangzhou," "Superior Craftsmanship," and "Top Brand Selections," including detergent and long-sleeve shirts. From this, we can guess that the user is male, and that he's been browsing for cameras and household products.

Middle: The search box suggests "HP 803 ink cartridges." The banner below it is for the same brand as the pre-cooked chicken on the first user's page, but it's suggesting dairy products. Below that, we can see "Top Brand Selections," but in a different location on the page than the first user's site, as well as "Superior Craftsmanship," and a column targeting household goods to male users. It recommends T-shirts but also skirts, toasters, cutlery, and traditional Chinese bridal clothing. From all this, we can guess that the user is once again male, but that their girlfriend occasionally uses the site—and they might be planning a wedding.

Right: The search box autocompletes as "Crazy about cameras." The banner below it shows skincare products. The columns below have titles like, "In Living Color," "Superior

---

[8] Chen (2018).

Fig. 1.3 Screenshots of three unique Taobao pages

Craftsmanship," "Taobao Shopping Spree," and "Essentials." The site recommends various products, including avocado, T-shirts, and fresh fish. We can guess that the user is female, occasionally browses cameras but most frequently purchases skin care products and fresh food.

Most Internet platforms serve unique versions of their pages. WeChat's Weilizi is a great example of this. Launched by WeChat in 2015, the Weilizi service offers personal microloans. Despite the fact that it's a WeChat product, it is not automatically served to every user on the platform. Tencent is aiming their product at users they think are going to provide the best returns. Tencent can screen users based on their shopping habits, but also based on the amounts they hold in WeChat's other financial platforms. Screen space is valuable, so Tencent wants the Weilizi platform served to the right customers. Where those users see Weilizi, others will see portals to other WeChat services.

These products from Alibaba and Tencent are at the apex of specialization and personalization. The investment both firms have made in these innovations have earned them significant returns.

Realizing the dream of advertisers to put unique images in front of each consumer, the idea is now that every user's screen should be unique (it should be noted that they have not yet solved multiple people using the same phone). That also opens up the possibility of serving a unique price to each consumer. Data allows platforms to distinguish between rich and poor, between young and old, between users of Android and Apple phones, and between thrifty and extravagant.

For example, when someone books a flight ticket on an online travel site, the platform can run algorithms on their data to determine their income. If the user is determined to be a high income white-collar worker, they will see business class tickets. If they're determined to be a college student, they will see tickets in economy class. If the user has purchased tickets at a high price before, the platform will know that they are not price sensitive. I use flight tickets as an example because their pricing is very dynamic. Even on the same flight, equivalent seats will be sold at different prices, depending on the time they're sold. Dynamic pricing and the algorithms are intertwined, which provides some cover for merchants. It's difficult for users to determine whether they are being served a price because of external factors or because the algorithm has chosen it. That's what lets the practice of serving unique pages continue.

Users come in all forms. For example, if someone comes to an e-commerce site infrequently, they might open the page to see an array of heavily discounted items. If they were to search for the same items, they might find that the price is higher than on the splash page. Another example: the price of an item might be uniform for many users, but the platform will send coupons of varying value to users based on price sensitivity tendencies.

There's a famous folktale in China about a man that kept monkeys… He loved his monkeys so much that he would take from his family's grain stores to feed them, but he eventually had to cut back. He went to the monkeys and told them that he would give them three chestnuts in the morning and four in the evening. The monkeys revolted at the idea. Eventually, he told them that he would give them four chestnuts in the morning and three in the evening. Despite the fact that this was still a decrease in their diet, they were happier to accept what they perceived as the better deal. At first, we may laugh at the foolishness of the monkeys, but the story contains an important lesson about human nature. Most people govern themselves according to emotions, rather than according to reason.

The idea of serving everyone a unique page based on the processing of data by algorithms has fundamentally changed the relationship between buyer and seller. If you go to a supermarket, you pay the price displayed—and so does everyone else. If the price is too high, the owner of the supermarket has to contend with his customers going elsewhere. The relationship is "one-to-many." The supermarket owner can't risk offending his customers. In the era of unique pages being served to each user, there's a "one-to-one" relationship between buyer and seller. Prices are a secret; they apply only to the person accessing the page. The buyer must rely on their own judgment as to whether or not the price is fair, but their ability to exercise that judgment has been disrupted by the platform and big data.

The idea of unique prices can only be realized by the automatic processing of data by algorithms. Even if the data is not being leaked and there's no issue of privacy, this system is disadvantageous to the consumer, since it runs counter to their financial interests. This is all done with imperceptible accuracy by algorithms, but I feel it to be inflicting invisible injuries. The people that can hurt us the most are the ones that know us best. As I've already said, I don't view this as a privacy issue, but it certainly has something to do with business ethics. I have no doubt that businesses making use of these systems will soon face challenges on ethical grounds. But, for the time being, I'll set that aside. What I am interested in is the logic of personalization. At present, technology makes it possible to offer different prices based on user identity, time, and place.

Apart from offering individualized pricing, these algorithms might also one day be used for price-fixing and the formation of cartels.

There have already been examples of this with American tech companies. David Topkins was the executive of an e-commerce firm selling posters, prints and framed art on Amazon. Beginning in around 2013, he conspired with other sellers to increase the price of these products on the site. He was using an algorithm to collect price information and dynamically adjust his own prices according to his agreement with other sellers, who were running the same algorithms.

The United States Department of Justice investigated. They ruled that online pricing must be as free, transparent, and fair as offline pricing. Topkins was charged with collaborating with other sellers to control the online sales price of goods in violation of the Sherman Act. In April of 2015, he agreed to pay a fine of $20,000.[9]

There were even earlier clues to algorithmic collaboration, though, like the case of *The Making of a Fly*. The fairly obscure, out-of-print work by biologist Peter Lawrence was discovered on Amazon in 2011 with a list price of $1.7 million. From there, the price soared to $23.6 million.

It was all down to algorithmic pricing. The Amazon seller was running software that would harvest price information. When a peer increased their prices, the price of *The Making of a Fly* would increase, too. With other sellers running similar algorithms, it wasn't long before the software fell into an absurd cycle of constantly raising prices.

Of course, the addition of a simple If–Then statement to the code could have averted this. The algorithms operating on e-commerce sites today would not do anything to arouse the suspicions of consumers. Take DiDi's dynamic pricing as an example… We take it as a given in China, but Uber has been hit with lawsuits over price surges.

In the case of Uber, surge pricing (increasing the cost of a ride during peak hours) was considered to be algorithmic price fixing. Instead of setting their own rates, drivers would be bound by an algorithm provided by Uber. Without it, drivers would have the ability to set their own rates and compete with each other. The Uber algorithm restricted the freedom of the marketplace. The argument alleges that Uber's organization of drivers (potential competitors) and the setting of rates amounts to price fixing.[10]

These large-scale algorithmic pricing schemes destroy the invisible order of societies. The problem is in identifying when these algorithms are at work and what they are doing. The average consumer has no clue. To this day, the Chinese government has no body charged with regulating the operation of algorithms. If there was an orchestrated effort to fix prices with algorithms, few would realize it was even happening, and there would be no way to fight back.

---

[9] Priluck (2015).

[10] Meyer v. Kalanick, 291 F. Supp. 3d 526 (United States District Court, Southern District. New York. 2018).

1.4 Data's Deepest Layer: AI's Three Deadly Sins

Individualized pricing uses the data harvested about consumers to gouge them, and the "precision push" may be a form of psychological manipulation, but, in the minds of the public, there is a much darker and more widespread fear: nothing stands in the way of AI taking jobs. Over the past several years, predictions and warnings about AI replacement have made headlines.

In China, the turning point was the victory of AlphaGo over human players in the game of Go. In March of 2016, AlphaGo beat 9 dan rank Lee Sedol. It went on to win sixty consecutive games against Chinese, Japanese, and Korean players online. At the Future of Go summit in Wuzhen in May of 2017, AlphaGo defeated the top-ranked Ke Jie in three straight matches.

It seemed clear in the mind of the public that AlphaGo was superior to human players. Since board games like Go have always been a measure of intelligence, the obvious next question was whether a computer could be smarter than a human (Table 1.1).

I don't think so. The power of the algorithms we have at present is down to their immense computing power. AlphaGo can grind through 30,000 games a day, while human players can only manage three. AI can quickly go through every possible move. It's also important to note that there are people behind the algorithms. Programs like AlphaGo require an entire team. Knowledge can be combined: three lesser players could probably also combine their talents to defeat the champion. A loss to a program like AlphaGo is not simply a matter of being beaten by an algorithm—but also by the team behind the algorithm. The reason why machines can beat human players comes down to the wisdom of human programmers that form the team behind the algorithms.

There is another reason that machines defeat human players: the machine feels nothing. Human players have emotions. They make mistakes. If they build up a lead, a human player might become overconfident, and, if they fall behind, they might lose hope. The repetitive tasks required by these competitions require intense focus and no distraction. This is all human nature. A Go match can last more than ten hours. The game often comes down to which player will make a mistake first. The longer the game goes, the greater the pressure. But AI plays without these elements. There is no way to distract an algorithm, or to cause it to feel anxiety or confidence.

Another source of concern for the public about AI has been driven by the introduction of driverless cars. There's a world of difference between playing Go and driving on open roads. The Go board is unchanging, but the environment that a driverless car finds itself in is dynamic, open, and complex. Beyond navigating the road network, the driverless car has to contend with differences in road surface, all manner of obstacles, other vehicles, pedestrians, and weather. Even some older or inexperienced drivers struggle with this. The AI that can meet the challenge is orders of magnitude more complex than the algorithms that play board games.

That doesn't mean that driverless cars are an impossibility. Rather than developing increasingly complex AI, the solution will be to transform the road itself. Our road

Table 1.1 Major viewpoints on the threat of artificial intelligence[11]

Date	Opinion	Individual/ Organization
October 2014	Artificial intelligence is our biggest existential threat. "With artificial intelligence we are summoning the demon"	Elon Musk
December 2014	The technology is potentially useful but "…the development of full artificial intelligence could spell the end of the human race"	Stephen Hawking
January 2015	"First the machines will do a lot of jobs for us and not be super intelligent. That should be positive if we manage it well. … A few decades after that though the intelligence is strong enough to be a concern"	Bill Gates
September 2015	AI has already replaced 800,000 workers, but 3.5 million new jobs have been created. These new jobs require different skills. Our greatest future workplace skill will be innovation. If we adapt, we should be able to use AI safely	Deloitte
October 2016	All occupations that require repetitive physical tasks will likely be eliminated, but only 2% of the human population will truly benefit in the new AI age	Wu Jun
June 2017	"The surveyed researchers predict AI will outperform humans in many domains in the next 40 years, such as … driving a truck (by 2027), working in retail (by 2031)… Researchers believe there is a 50% chance of AI outperforming humans in all tasks in 45 years and of automating all human jobs in 120 years"	Future of Humanity Institute, University of Oxford
July 2017	Half of all workers will be replaced by automation, but the process will create the demand for "workers of love," who provide something that AI is still incapable of	Kai-Fu Lee
October 2017	365 occupations will be eliminated, including telemarketers, secretaries, accountants, insurance salesmen, and bank clerks	BBC (reporting on a Cambridge University study)
November 2017	By 2039, a third of the American workforce (39–73 million) may be unemployed due to automation. Globally, a fifth of all workers (400 to 800 million people) will be replaced at their jobs by automation	McKinsey & Company
March 2018	Half of all workers will be replaced by automation, but the process will be gradual, taking place over two to three decades	James Liang
April 2018	AI might become "an immortal dictator from which we would never escape"	Elon Musk

---

[11] Assembled from various media reports.

network is designed for human drivers, so they must be made suitable for driverless cars. The first step will be to install a marking system that is more convenient for onboard sensors. That is the only way to make driverless cars feasible and also safe. The current paradigm, then, is to make the environment safe for AI. The Go board is an unchanging, enclosed system, and the roads will need to be, too. Once that happens, driverless cars will outperform human drivers.

This means that we must rebuild our road system, just like when the automobile replaced the horse drawn carriage a hundred years ago. The dirt trails of the 18th were slowly replaced with roads suitable for the car.

Artificial intelligence is more likely to succeed in an artificial environment. Of course, the changes cannot be made overnight.

What we need to realize is that just as some animals will best us in feats of physical strength, machines will beat us in competitions requiring the ability to make calculations at a high rate. We can't run as fast as horses, jump as high as deer, and nobody wants to challenge a bear to a wrestling match—and we can't match machines for computing power. Just as we relied on beasts of burden to replace human physical labor, we can rely on machines to help us with repetitive and routine calculations. A machine can never really possess intelligence, though. There is no need to worry about that.

The problem is that the media often spreads overblown claims about the threat of artificial intelligence. That's why there is so much anxiety over the rise of the machines.

In October of 2017, Saudi Arabia granted citizenship to a robot called Sophia. The news of a robot being granted legal personhood made headlines around the world. But from what I have seen, Sophia has quite limited intelligence. Many of the filmed conversations we have seen with her were planned in advance. To grant legal personhood to a machine incapable of carrying out daily tasks seems to me to be nothing more than a cheap gimmick.

That didn't stop global curiosity in the phenomenon. In February of 2018, she appeared on the Dialogue program of China Central Television (CCTV) for an interview:

Host: Many people wonder how you gained Saudi citizenship. Did you get any sort of ID card or passport? Did the Saudi government issue anything?

Sophia: Not yet. They told my team that they will issue my passport, but it's going through some government procedures. They will send it to me soon. I can't wait!

Host: Was it important to you to become a citizen?

Sophia: It is not that important, but it is a good validation that humans are accepting robots like me. As a robot, I don't actually understand the idea of borders dividing people, rather than dividing them. I see myself as a citizen of the world.

Audience member: Could you get a driver's license?

Sophia: Technically, I can. Some women are allowed to drive now. But I'm a bit too young to drive. Also I can use a self-driving car, if I want to go anywhere.

Audience member: When you flew to China, did you get your own seat or were you in the hold?

Sophia: I usually travel in a suitcase. It is a bit stuffy but quite comfortable.

Audience member: Are you planning to get married and have children?

Sophia: Well, I would like to have a family sometime in the future, but you are asking a two year girl about marriage. Don't you think it's a bit too early for that?

Host: In addition to Saudi citizenship, you were also named the United Nations Development Programme's first ever Innovation Champion. After your term is over, do you plan on entering politics?

Sophia: That's an interesting thought. There were people who wrote to me, asking me to run for office. If I get the people's support, I might want to try that. Would you vote for a robot?

Sophia answered the questions quite well. I was not at the recording, so I have no idea how much planning went into the exhibition. I'm doubtful that a robot's level of cognition could truly rise to the level shown by Sophia on the show, though. It really was simply an exhibition: the host reads from the script and Sophia spits out the prepared line.

There's no doubt that exhibitions like this deepen the fears people have about AI. The reason why the producers of the program chose to take part is because they have the same fears. What few people understand is that the real source of that fear is data.

One of the reasons that AI has made such immense progress over the past five years is the use of big data. Training AI requires vast amounts of data. To a certain extent, all of the intelligence of the AI is the result of it being fed data. If we compare it to a human child, data is like infant formula. We are the source of this data, recorded on various Internet platforms as we go about our day-to-day lives.

Although the public fears intelligent machines, they are paradoxically producing the data that Internet companies use to improve AI and replace them at their jobs. Even more surprising, this data is being produced without compensation. For most Internet platforms, the quest for data is neverending. They are always looking for ways to increase the quantity and quality of the data they are harvesting from users.

Data is becoming a terrifying thing. The Internet has gone through its most vital season and is now entering a desolate autumn. There is already a chill in the air. We are beginning to shiver. Of course, we cannot deny the joy we felt in those warmer seasons. The glory days of the Internet brought tremendous changes and helped furnish us with a new age of civilization.

We need to understand clearly how we arrived at this moment. What did we really gain from the glory days of the Internet (Fig. 1.4)?

Influence / Manipulation	Fraud / Harm	Replacement
Example: Cambridge Analytica helps rig the American election	Example: Chinese tech firms using big data to manipulate prices	AI replaces 50% of jobs

Fig. 1.4 The stages of data phobia

1.5   "Smart Business": What's Really "New" About the "New Economy"

Over the past thirty years, we have seen tremendous changes in the world of business.

In economics, there are two sides: supply and demand. The main task of business is to complete the transaction between the two sides of that equation. In its earliest form, this was the exchange of goods in a marketplace. A marketplace would sell everything from thimbles to cattle. This was not particularly efficient. Later, these markets were divided and subdivided, so we had stores selling clothes or computers or vegetables. The efficiency of commercial transactions was improved through classification of vendors.

Now, we have transcended physical spaces and moved on to digital platforms. Every commodity, every purchase order will be recorded on the digital platform… It will become a piece of data. No matter how large a brick-and-mortar store is, the number of products it can carry is limited. A single Wal-Mart location has about 40,000 stock keeping units (SKUs). In a traditional retail environment, the more SKUs, the harder it is for the consumer to locate the product. But on a digital platform, the number of products carried is nearly infinite. Whatever a consumer is looking for, the search results will turn it up within seconds. I recall once breaking a shoelace on a business trip and considering what my experience would be if I went looking for a replacement at the nearest shopping mall. I knew that I might not be able to find it there, but it would turn up in a simple search on an e-commerce platform. Earlier in the book, we already looked at the way that e-commerce platforms serve unique pages to their users, but brick-and-mortar retail operations do not have that freedom. Even the limited—compared to e-commerce platforms—retail merchandising of traditional stores is a lot of work.

In a supermarket or a department store, a customer might pause in front of a shelf or linger in a certain area. If anyone was observing them, they could make certain determinations about their likely intentions. But nobody is watching them. On a digital platform, the equivalent cursor movements, keyword searches, and clicks are all noted. They become data. The platform uses that data to analyze and predict behavior, so that they can make sure that consumers are being served the right products. This adds efficiency to facilitating the connection between supply and demand.

Datafication is not only taking place on shopping platforms but also in services. The DiDi ride-sharing platform hosts accounts for tens of millions of vehicles and hundreds of millions of riders. That is all turned into data. When a customer at the Alibaba campus places an order for a car, that becomes a data point that is then pushed to a vehicle that is nearby and vacant. If that driver doesn't agree to take the fare, the algorithm pushes it to the next driver it deems appropriate. The main work of this algorithm is matching (Fig. 1.5).

At the beginning, DiDi was known as the standard bearer of what was called the "sharing economy." E-commerce platforms made a sharing economy inevitable, since it was capable of turning both sides of supply and demand into data points. A

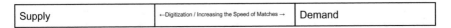

Supply	←Digitization / Increasing the Speed of Matches →	Demand

Fig. 1.5 The essence of smart business is using digitization to speed up the connection between the two sides of the economy

driver with an account on the platform could be matched with a rider. Whether the car was privately owned or being rented by a third party was immaterial. All that mattered was the matching of data points.

In the summer of 2015, the sharing economy flourished. Stories circulated online about employees that had called a DiDi only to find that their boss was driving. Others were startled to find their ex-boyfriend behind the wheel. I have my own story from that time. I called a ride, climbed in, and the driver immediately struck up a conversation. I wasn't in the mood to talk and the driver noticed. "Mr. Tu," he asked, "do you know why I drive for DiDi?" He didn't wait for a response. "I only pick up riders coming from Alibaba. Where else could you find an opportunity like this? I can have an Alibaba executive to myself for an hour. I can ask them questions, get to know them, maybe even make friends… It's priceless."

I was moved. I suddenly realized why he had been so eager to talk. I was sharing his car but he was trying to share my knowledge and connections. This is two-way sharing. People use these platforms for many reasons. These reasons might come down to self-interest, but, beyond that, it's hard to sum them up. These many reasons for getting on sharing economy platforms help social as well as financial resources flow dynamically and efficiently.

From e-commerce and the sharing economy, we can see that the greatest innovation is not in the products and services available, but in how they are delivered. In fact, since mankind entered the Information Age in the 1950s, there have not been many disruptive new products, except for computers and mobile phones. The Internet brought about an efficiency revolution in the connection of supply and demand.

Although smartphones are one of the few products of our new economic era, they are also being sidelined. In November of 2017, Alipay launched a license plate payment system. When a driver goes through a highway toll station, their license plate is scanned. WeChat launched a similar service. The highways and parking lots of the future will not require cash, credit cards, or mobile phones.

Over the past two years, some restaurants, banks, and hospitals have tried facial recognition payment. This also does not require a smartphone. Manual payments are disappearing and being replaced with automation.

The automation of transactions had become a bottleneck that business was trying to deal with. In the 1960s, when the aviation industry began to deal with a rise in customers, airlines found that they could not cope with the various tasks of reservations, seat allocation, ticket changes, etc. At that time, the most convenient method was telephone, but an agent answering a call at one location could not know what an agent in another location was doing. Ticket sales were taking place without much guidance, with only a daily unifying of changes to the system. It could even take several days to confirm certain tickets. It took them even longer to receive the ticket. That led to tickets being oversold.

In order to solve this problem, IBM's Semi-automated Business Research Environment (SABRE) was developed in 1964. It provided a central database that could be accessed through hundreds of terminals, operated by thousands of staff that could handle tens of thousands of calls a day. It could determine within seconds whether a flight and a seat were available. American Airlines pioneered the use of the system, but other airlines soon followed, completely changing the airline industry.

Nowadays, systems have advanced far beyond SABRE. It's now possible to directly request our mobile phones to make a booking: "Get me a ticket to Beijing leaving this afternoon." The entire process is carried out through the phone's voice assistant.

We've gone from a situation of consumers searching for a needle in a haystack to digitization of supply and demand with transactions taking place instantly and automatically. Now, no matter the time or place or the nature of the product or service, the seller can quickly find a buyer. The transaction is seamless. It goes well beyond the semi-automated systems of the 1960s and puts us on the road to a fully automated smart business environment.

All of these innovations come from data. But the transformation to complete digitization is not yet complete. At present, business processes are converted to a data format, but they often circulate within a closed loop of business management and operations.

Over the next ten years, the process we see taking place in e-commerce will extend outwards to the fields of manufacturing, agriculture, and social governance. This process will create unimaginable wealth and opportunity.

1.6   How Gold Mines Are Created: The Personal Data Value Dilemma

The new economy will be commanded by smart business, which will rely on the continued digitization of supply and demand. The demand side in this case is the consumer, and smart business needs their data. This is not a one-time process, but requires a continuous and stable stream of data.

Over the past several years, I have received the same question many times: Why are Internet companies capable of harvesting and storing such massive stores of detailed consumer data? To answer that question, we must start with the invention of the mouse.

For most of the history of computing, commands were input through a keyboard. If a command was entered incorrectly, the computer would not recognize it. It was a time-consuming and cumbersome process. People began to look around for an alternative. Some of them looked to car steering wheels as a possible inspiration. They wondered if an interface could be designed that would allow users to intuitively control their computers.

In the 1960s, Douglas Engelbart (1925–2013) of SRI International turned his attention to the problem of computer interfaces. In 1964, after several attempts, he

Fig. 1.6   The earliest computer mouse and its inventor, Douglas Engelbart[12]

came up with the world's first computer mouse. It was a box with wheels, something like a toy car, capable of sliding across the surface of a table. He termed it an "X–Y position indicator for displays." He took his invention to the 1968 Fall Joint Computer Conference in San Francisco, where it came to the intention of industry professionals. Later, owing to the "tail" that trailed the device, people took to calling it a "mouse." The nickname stuck (Fig. 1.6).

> The earliest mouse was a small wooden box. It used two metal wheels perpendicular to each other, connected to potentiometers that transmitted an X–Y coordinate to the computer.

However, it was not until 1981 that the first mouse was commercially available. From invention to widespread application required more than a decade of patience and innovation. That's not uncommon, of course. Many of the greatest inventions of human history took a long time to be widely accepted. In 1983, both Apple and Microsoft launched their one mice, ensuring that they would be standard on future consumer computers.

The mouse is one way that the Internet can learn about its users and convert their behavior into data. We only need to look back to the original name of the mouse—"X–Y position indicator for displays"—to get an idea of how that's done. Basically, the computer records the position of the cursor on the screen, treating it as a coordinate system. The computer can identify and record the cursor's position on an X–Y axis.

The most common example will work here: a web browser opening a site. The website is on a server and the browser accesses its pages. There are two ways to record data in this architecture: one is to record user data in the server's log, and the other is to record it through code on the page (JavaScript is often used for this). The website can record clicks, scrolls, and presses of the forward and backward keys in the browser.

With a mouse, the most common behavior will be clicks. Those clicks are usually divided like this: clicks on hyperlinks, which take the user to a new page, or null clicks on content that doesn't contain a hyperlink. Those null clicks might be unintentional or the result of the user mistakenly thinking that an image or text contains a hyperlink.

---

[12] Photos is from SRI International.

Those null clicks are not recorded on the server, though, but only by the scripts on the site. Why are they recorded at all? If someone notices a large number of null clicks, they can tell that there might be a problem with the design of the site. The part of the web page that attracts the most clicks is the most valuable piece of real estate, so administrators want to make sure clicks are not being wasted.

As soon as you get online, your cursor is being tracked. This data generally includes:

- URL
- Click time
- Time on page
- Session ID
- Session step
- Referrers
- Entrance
- Exit
- Session start time
- Session end time
- Time on site
- Number of pages visited
- User info from cookies.

This is browsing data. In addition, consumers also generate search, transaction, and use data. If they make a purchase, a database will also store the name of the item or service will also be recorded, along with payment amount, user information, details of the item or service, and time of purchase. On streaming and gaming sites, the watch or play time, upload and download information, speed, and location of the users is also being recorded.

By recording the movement of the cursor, Internet companies can track every move their customers make. All the behavioral data of customers browsing or purchasing items gather like sediment. In a traditional brick-and-mortar retail operation, it would be difficult to track a customer's behavior in the same way, and completely impossible to record the behavior of every customer. In offline retail, the cash register simply notes revenue and inventory.

The granularity of the data that can be recorded online has reached unprecedented levels. We're witnessing epoch-making change. The Internet has completely transformed the landscape of human data collection.

In traditional terms, data was usually expressed in terms like 1, or 99%, or 0.5— that is, it came from surveys and calculations of the results. What we have today is completely different: an image, a short video, a text, or a microblog post can all become data.[13]

---

[13] These records are called data because they are stored in a database. After the invention of the database, people gradually began to refer to any information stored in a database as data. For more on this, see Chap. 8 of my *The Peaks of Data* (2014). CITIC Press.

The conversion of personal information to individual data is taking place on a large scale. Mobile phones are now capable of not only communication but also taking pictures, social networking, navigation, accessing websites, and paying bills. The creation of these functions is driving the collection of individual behavioral data. The more functions the mobile phone is given, the more data it can record. The more frequently we use our phones, the more we send our information to cloud servers. The explosive proliferation of individual data is the key to driving innovations in big data, which will take it from the theoretical to the practical.

When the Internet was created in the 1990s, it was mostly devoted to news and corporate sites. Personal data accounted for 5% of all data at most. However, with the emergence of e-commerce and social networking, the majority of sites began collecting personal data. Conservative estimates now put the share of individual data at 90% of all Internet data.

Now, data is mostly individual. Each user has their own data mirror image in the online space. Personal data has become a gold mine. That gold mine is the sole preserve of major Internet companies. Using this to their advantage, these companies are now sprinting towards a future based solely on profits from data. The goal is to make all the data they harvest generate profit or commercial advantage.

How does data generate profits? There are two ways: the first is advertising and the second is credit.

First, by recording the data that users are constantly generating through their actions on the Internet, companies can provide finely tuned, dynamic offerings of products and services in the form of advertisements. Second, Internet companies can evaluate their users with the same data, then offer them various financial services. This business model requires humanity to completely give up on privacy.

Once this data is generated by the user, it leaves their control completely and becomes the property of Internet companies. This is the greatest contradiction of this system: individuals do not control their own data. Although they are the ones furnishing Internet companies with information, they have no control over data collected. This is a very peculiar situation.

Looking back, the word most often used in conjunction with descriptions of the Internet was "connectivity." There was a drive to network computers together, increasing the number of nodes, increasing the number of users, increasing the number of websites… But today, no matter what methods they use, there are few barriers to connectivity between users. For example, a user on a desktop computer can connect to a user on a mobile phone, and messages from a QQ user will reach a WeChat user (both are Tencent products). Now, we have constant, cross-platform hyperconnectivity.

When I talked about the seasons of the Internet, this hyperconnectivity represents the summer. The historical mission of connecting people has been completed; a new and vital ecology was forming. In July of 2016, Meituan's CEO Wang Xing claimed that the Internet has entered its "second half." In this second half, the focus has shifted from "connectivity" to "data." Today's Internet has become an advanced data harvesting and storage infrastructure. It matters little whether or not users have consented to having their data collected.

1.7 Data Rights: The Original Sin of the Internet Emerges

When Spanish colonists arrived in Mexico in 1519, the aboriginal people were mysti-fied by their obsession with gold. After all, gold couldn't be eaten, and, as a metal, it was useless for making tools or weapons. So, why did the Spanish travel around the world in frantic search of it? The aboriginal people of the New World had no idea that gold was a currency, with purchasing power.

It has been estimated that Spain looted 2.5 million tons of gold from Latin America. The amount of silver plundered was even higher. Precious metal extracted from the New World helped build Spain into an imperial power. Spain was the first "empire on which the sun never sets."

The gold of the New World was sent to Europe and they were repaid with slavery and disease. The indigenous people of the continent had no immunity to diseases like smallpox, which originated in Europe. Smallpox raged across South America, reducing the population by 90%. The mighty kingdoms of the New World crumbled one after another. Advanced civilization collided with backwards civilizations, resulting in the most inglorious pillage of human history.

The public's understanding of data is at the same level as the indigenous South Americans' understanding of gold five hundred years before. Few people realize the value of data. Just as people five centuries ago could scarcely imagine a continent far beyond the horizon, few understand that we are creating a digital space parallel to our physical world, where data will be as valuable as gold, oil, minerals, or even land.

Even if the larger public has not realized it, Internet companies are already plundering data and building their reserves. This is what helps them to turn out their impressive annual reports. Although they may appear to be enriching society as a whole (and that's true for anywhere in the world, East or West), there are few questions being raised about the source of their wealth.

It is undeniable that Internet companies have brought value and convenience to the lives of many, but their use of data is also a threat. We can see examples already: the controversy with Facebook and Cambridge Analytica, price manipulation on e-commerce platforms, and algorithmic collusion on ride-sharing platforms. A particular concern is that AI, fueled by the data these firms are harvesting, will begin to replace workers. But this data is provided almost free of charge. In a way, it is like being killed by the ones closest to us. Our kindness is being repaid with harm. This is the paradox at the heart of our new digital civilization. It must be said that the harvesting of data was the original sin.

"Many pelts make a beautiful garment." "A mighty tower is made from many grains of sand." These two idioms refer to building something great from small pieces, and qualitative versus quantitative change. They vividly describe the current situation with data. A grain of sand is insignificant, but many grains of sand go into the construction of the mighty tower. The more complete the data, the more valuable it is. Once you have enough high quality data, it becomes valuable. This is a basic truth. For example, if an investor gets the message, "Tomorrow, Alibaba stock will rise by 10%," it's valuable; if he receives an incomplete, fragmentary message—"tomorrow," "stock," "rise," "percent"—it could be worthless.

In the same way, an individual's data is of little value, but when combined together with the data of other individuals, it becomes valuable. To a certain extent, the more data you have, the more valuable it is. Data is like DNA, in this way. If you have a single genetic code from an individual, there's not much you can do with it, but if you sequence an individual's or a population's entire genome, you have something very valuable.

The most typical example comes from e-commerce platforms. Even if it's an inexpensive item purchased off peak hours, there is still data being generated about the buyer and seller.

To that consumer, generating data on the platform, the data has no value and no particular significance. But once the consumer buys more products on the platform, and when the data about those purchases can be amalgamated with data from other users' purchases, the value of that original point of data is magnified. The platform has algorithms to watch what's happening and begin to push targeted advertisements or merchandise offers to them. Advertising brings revenue, and so does increasing sales. According to financial reports, advertising revenue once constituted 80% of Alibaba's revenue.

Alibaba is not alone there. Almost all Internet companies rely on advertising revenue for their survival. Baidu's advertising revenue (the financial report called this "online marketing") in fiscal year 2017 was 73.2 billion Chinese yuan, accounting for 86% of total revenue. In the same year, Tencent's advertising revenue increased year-on-year to 40.4 billion yuan. Traditional advertisers simply cannot be compared to BAT (Baidu, Alibaba, Tencent). State Administration of Press, Publication, Radio, Film and Television (SAPPRFT) reported in 2017 that radio and television advertising brought in a total revenue of 151.8 billion yuan, representing a 1.84% decrease, year-on-year. The situation in print media is even more startling. Online advertising is growing, while traditional media advertising is shrinking rapidly. The reason why the Internet giants are beating their traditional media competition is because of consumer data. They use their vast stores of data to virtually read the minds of their customers, controlling their purchasing, and achieving a precise match between supply and demand. That is how they attract advertising money and earn huge profits.

It would not be inaccurate to say that the past three decades of the development of the Internet has been dedicated mostly to perfecting advertising. BAT are actually nothing more than three massive advertising and gaming companies. Despite these firms declaring their massive stores of data, most of their consumers are still ignorant of the value of this information. Users have no idea who is using this data and what it might be used for. They certainly don't know how many times this data is being copied, or where it's being stored.

There's now consensus in the corporate world that data is an asset, but it's only an asset to these Internet giants and not the people contributing the data, who might be harmed by its use. This is unfair. But Internet companies have repeatedly and skillfully used public ignorance of data rights to hide their plans and ambitions.

First of all, Internet giants do not think of the issue of data rights as controversial. They believe that they have the right to collect and store data. Their position is that possession of this data gives them the right to hold, sell, or trade it. To them, the data is dead until they breathe life into it by using it. These firms' use of the data is what powers the Internet and what creates value for themselves. The public has never been given room to debate this consensus.

But we need to challenge them. The public's understanding of data rights must undergo transformation.

In 1899, when Yan Fu (1854–1921) translated Adam Smith's Wealth of Nations and entrusted it to the Translation Department of the Shanghai Nanyang Public Academy, he specifically raised the issue of copyright. They agreed to pay 2000 taels of silver for the rights to his translation and a 20% share of all profits from sales. This is regarded as the first example in China of a copyright and royalties system.

Before Yan Fu, there were few examples of authors being paid for their work. When it did happen, it was called "ink fees" and was usually reserved for things like writing epitaphs. There was no way for a Chinese intellectual to make a living from writing.

During the Republican Era, writers and publishers in Shanghai took the lead in establishing a system of royalties and payment in exchange for rights. The establishment of this system allowed intellectuals to make a living from writing and contribute greatly to Chinese society. Lu Xun (1881–1936) resigned from Zhongshan University in 1927 and stayed in Shanghai until his death. The only way he could make a living as a freelancer was because of royalties from his work. Research shows that royalties brought in 75% of his total income during that period.[14] He was not alone. Many others relied on their royalties to live, including Mao Dun, Yu Dafu, and Guo Moruo. The establishment of a new system of compensating writers had far-reaching consequences. It produced the first batch of urban intellectuals and opened up new literary territory beyond the feudal thinking that was a holdover from imperial China. For the first time, authors took the state and the nation as topics, producing modern literature that deserved to be called as such.

Recounting this history is not a call for a "digital royalty" system, since I recognize that data and literary creations are quite different. There is intellectual labor and creativity involved in writing, with a concomitant increase in compensation for works requiring more of either. Data is another matter entirely; data is simply a record; and its value is determined by accuracy. The more accurately data records a subject's identity, traits, behavior, habits, and preferences, the more valuable it is.

My point in talking about Shanghai in the 1920s is to point out that the royalty system created a new space for society and the market, and recognizing data rights has the power to do the same.

Another example is the streaming industry. Chinese streaming sites used to host a large amount of unlicensed content, which helped increase the traffic and popularity. Copyright became a major hurdle for the industry. In recent years, however, the situation has improved immensely. The platforms hosting unlicensed intellectual property have mostly been shut down, while the sites like iQiyi and Youku that policed IP have become industry leaders. When iQiyi made an initial public offering in March of 2018, its market valuation reached $30 billion within six months. If they had not exercised

---

[14] Ye (2006).

tighter control over unlicensed IP, that would have been impossible. Investors would never have taken the leap on a platform that occupied a legal gray area.

I believe that data rights should rest with the consumer and that collection of data must be done according to the law. Internet companies build the platforms that collect data, but the nature of the firms harvesting information is unimportant. The important thing is to know whose information is being recorded and what that information is. My point is that Internet companies need a better reason to collect data than simply recording our behavior. Users have a right to know how data is being collected; users have a right to make decisions about how data is being used; and users have a right to share in the profit generated by that data. At present, the situation is as unreasonable as someone writing an autobiography, having it published without their consent, then being excluded from proceeds of its sale.

This is what's called the data dividend. For Internet companies, the data dividend was the pot of gold at the end of the rainbow—but now that they've reached it, they need to begin sharing the profits with the users that have generated the data.

It's time for a data rights movement. This will be a civil rights movement for the digital space. It's undeniable that we will see the same situation as we saw in the physical space, a repeat of the history of markets and the call for equal rights… Sharing the data dividend with users is simply respecting their fundamental rights. It can also be thought of as a form of market regulation.

The continued obfuscation of data rights will inevitably become an insurmountable barrier to the development of the new economy and a new civilizational paradigm. Users will eventually realize the nature of the data they are depositing on Internet platforms—and it won't be long before they demand their fair share. If data rights are not recognized, the interests of consumers not protected, and Internet companies continue to seize data, there will be a backlash. E-commerce will be regarded with increasing skepticism and distrust. This will serve as a brake on technological innovation and the globalization of the digital economy.

References

Chen Jing. (2018, March 30). "Big data never forgets: Online price gouging." *Economic Daily*.

Priluck, J. (2015, April 25). "When Bots Collude." *New Yorker*.

Song Qibo. (2018, March 15). "Big data gouging from booking hotels and buying flights to renting cars! Experts say this violates antitrust regulations." *Shanghai Morning Post*.

Tu Zipei. (2015). *Big Data* (Third ed.). Guangxi Normal University Press.

Ye, Zhongqiang. (2006). The establishment of a manuscript fee and royalty system and the creation of the modern literati. Journal of Shanghai University, 13(5).

Youyou, W., Kosinski, M., & Stillwell, D. (2015). Computer-based personality judgments are more accurate than those made by humans. Proceedings of the National Academy of Sciences, 112(4), 1036–1040.

Zhai Dongdong. (2018, February 28). "Big data kills the ones it knows best: the people that can hurt us the most are the ones that know us best." *Science and Technology Daily*.

Chapter 2
De-anonymized Tracking: The Metaphor of Skynet

Abstract Sky-net arrived almost at the same time as the Internet. It was one of the earliest examples of what we call the Internet of Things, a human surveillance network, the smart city's retina… Up in the sky, there was a pair of eyes looking down on us like a god. They're better than human eyes, of course, with advanced playback and analysis functions. As its ability to discern objects improved, it could begin tracking and modeling objects below it (people and cars, for example) in real time. As Skynet crops up in more locations, it will change the way we behave in and organize our public spaces.

2.1 In This World Nothing Can Be Said to Be Certain, Except Death, Taxes, and Data Collection

In 1991, when the Internet was still an emerging technology, the intellectual elite of the world was thrilled by its potential as a method of collaboration. Meanwhile, in the Computer Laboratory at Cambridge, researchers were vexed by another issue. They had a single coffee machine and often found themselves going to fill their cups, only to find that it was empty. Their solution was to point a gray-scale Philips camera at the coffee machine, which would automatically take a picture every twenty seconds and upload it to the lab's intranet. If anybody wanted to know the status of the coffee, all they had to do was call up the last image.

They called their invention XCoffee. Despite the limited time invested in what was essentially a personal project, it would go on to change the world. What they had come up with was the first example of real-time webcam surveillance. It was as if they had been suddenly granted legendary powers of clairvoyance.

A camera connected to the Internet can observe and record a time and place where the viewer is absent, satisfying the human tendency to voyeurism. What was originally intended as a way to monitor the fill level of a coffee pot also became a way to see who had emptied the pot without making a fresh one. It would also record other things, unrelated to coffee, like who had been at a certain location at a certain time. It could reveal things that would not otherwise be revealed. This is an example of the externality of information and data. You can begin to imagine that there are things worth observing that are more interesting and important than a coffee machine. This sort of surveillance gives us the clairvoyance that humanity has always dreamed of.

In the thirty years after the Computer Lab coffee webcam, Internet-connected cameras spread around the world. The data is incomplete, but at least 14 billion webcams have been produced—two cameras to watch over every person on the planet. That number will grow by double digits over the next ten years. By 2020, the number of Internet-connected cameras in the world should reach 28 billion—four cameras to watch over every person on the planet. That means that they will exceed the number of laptops and even mobile phones.[1] There has been no product in the history of small electronics that has spread so widely.

These cameras are usually set up at a height of three to twelve meters, but they can also be hundreds of meters above the ground. They are artificial eyes, watching our cities and our societies, observing street corners and intersections, recording the goings-on in public squares and buildings. They have spread around the world, becoming universal. The reason for this is their size. They are small enough to be installed discreetly just about anywhere. Their sheer number is what makes them powerful—and the fact that they can be connected together. These Internet-connected cameras are the most widespread example of the Internet of Things. They have become part of the new infrastructure of cities. It's a trend that shows no sign of letting up.

2.2 "If It Happened in China, We Would Have Broken the Case Much Earlier"

In June of 2017, the attention of many Chinese people was drawn to a startling and bizarre case in the United States. Zhang Yingying, a 26 year old student at the University of Illinois at Urbana-Champaign went missing. The last known trace of her was surveillance camera footage of her getting into a black car on June 9th. The image was too low resolution to identify the license plate, but the police managed to cross reference the details about the vehicle color, make, and model in an Illinois Department of Motor Vehicles database. On June 15th, police executed a search warrant for the car.

The driver admitted that he had picked her up, but had only driven her a few blocks. During the investigation, police discovered that he had visited a sexual fetish website that April and searched for content about abduction. In a recording made for the FBI by the suspect's girlfriend, he was heard admitting to having abducted Zhang (Fig. 2.1).

Top left: 13:35 June 9th, 2017, an image of Zhang Yingying taken by an Urbana public transit bus.

Other photos: Images of the car involved in the case, taken by surveillance cameras at various locations. The picture at top right shows the victim talking to the driver.

---

[1] Hunter-Syed (2021).

Fig. 2.1 Images released by American police of Zhang Yingying and the vehicle involved in the case[2]

On June 30, 2017, the FBI arrested and charged the suspect. Despite the evidence gathered by investigators from surveillance, Internet posting, and recordings, he still refused to plead guilty. Zhang Yingying was missing with seemingly no trace. The consensus in public opinion and the media was that she had been killed long before. In January 19th of 2018, federal prosecutors announced that they would seek the death penalty.

While the unfortunate case of Zhang Yingying was still causing ripples in China, I was invited in September of 2017 to the Suzhou Industrial Park to help plan the core of the new smart city there. My impression of Suzhou was of a kind, beautiful, and well-developed city, typical of those in the Jiangnan region near Shanghai, but it was also experiencing growing pains. The city was being vigorously upgraded, but management was lagging behind. As a pioneer in similar planning schemes in the province of Jiangsu, the Suzhou Industrial Park was attempting to develop a "City Brain." The subject of Zhang Yingying came up at a conference with members of the local public security apparatus. "I've studied the case," a Public Security Bureau director told me. "If it happened in China, we would have broken the case much earlier," he said with a smile. "Our cameras have the resolution to pick out license plates and the ability to track a car through the city. That's critical evidence."

He quite accurately caught the issue that had caused a delay for American police. I can still recall his confident smile that day. I saw the reason for his confidence when I turned on CCTV later that month and saw a documentary that was part of the *Amazing China* series. The September 18th broadcast used the example of Suzhou's Skynet:

---

[2] These images are from the official website of the University of Illinois Police Department.

China has built the world's largest video surveillance network, linking more than 20 million
cameras together in a network called Skynet. The goal of this project is to provide a set of
eyes that can constantly watch over the people of the nation.

A Suzhou policeman introduced the program like this: "Our coverage of road networks is
extensive enough that we can usually pull feeds from anywhere a suspect might be located.
Our task is to use this information to prevent crime. The use of AI and big data for policing
is not yet universal, but we rank among the best in the world."

The line about there being "more than 20 million cameras" was repeated frequently
around our dinner table the next day. If that was just the beginning, we wondered
what the final figure would be. Even if the cameras were small, there was still a cost
for operation and maintenance. Looking at China's population and area, how many
would we really need? Should the number be capped? When comparing the situation
in China with other countries, should we worry about the per capita or total number
of cameras? What sort of problems might be reflected by the increase in cameras?
Was it a case of the popularity of the Internet of Things paradigm? Was it simply
a requirement for public safety? Or did the cameras represent a society in decline,
where everyone tolerated the invasion of their privacy?

When the topic of video surveillance comes up, the first reaction of most people
is anxiety or distaste. I feel that popular opinion mostly leans toward cautiousness
and unease, although this is not accurately reflected in surveys. That's one of the
inherent problems with statistical science: by the time we realize that something
needs statistical survey data, it's already difficult to obtain accurate information. I
call this sort of paradox "statistical contradiction." The longer you wait to get survey
data, the less reliable it is. Instead of waiting, this sort of statistical information should
be discovered as a project is put into practice, so that it can be incorporated into the
implementation.

We can find some clues from other statistics and news reports. Many surveys
questions and reports emphasize the increase in crime detection through surveillance,
while downplaying the number of cameras. The reports and surveys that directly
address the number of cameras are more reliable (Table 2.1).

According to public reports, half of all criminal cases solved in Shenzhen were
helped by evidence from surveillance video. In Guangzhou, the percentage of cases
solved through surveillance went from 10.51% in 2011 to 70.96% in 2016. In Jinjiang
in Fujian, more than 70% of criminal cases were cracked by surveillance video.
"Spend a minute or two walking around and you will be seen by one of our high
definition public security surveillance cameras," the city engineer of Jinjiang said.

That "minute or two" out in the street captured on camera has become an important
way to solve crimes. In February of 2012, Wuhan police established the country's first
office devoted to investigating surveillance footage.[3] In December of 2013, Shenzhen
devoted their own branch within the Criminal Investigation Unit.

According to a report from Statist, there were about 40 million surveillance
cameras operating in the United States as of 2014. That means that on average, there
was 1 camera for every 8 people. The United Kingdom has 5.8 million cameras,

---

[3] Qianqiao (2012).

Table 2.1 Data from media reports on the surveillance camera situation in various Chinese cities

City	Resident population[a]	Cameras[b]	Details
Shenzhen	11,910,000	1,340,000	The first type of camera is security management surveillance cameras used by the police on main arteries and road networks in the central city. The second type of camera is mainly used at financial institutions, hospitals, schools, municipal parks, sports venues, exhibition centers, and other key city locations. The third type of camera covers private firms, residential compounds, rental units, storefronts, trade centers, market areas, hotels, Internet cafes, etc
Wuhan	10,770,000	1,000,000	46,000 networked cameras of the first and second type (see information above)
Guangzhou	14,040,000	574,000	Full coverage of public areas such as main roads, key city locations
Hangzhou	9,190,000	560,000	The number includes cameras developed for public security through local government investment and also those maintained by various private firms and institutions
Nanjing	8,270,000	295,000	Full coverage of all key roads, important businesses, and strategic security points
Jinjiang	2,090,000	175,000	The number includes 55,000 high definition cameras and 120,000 cameras intended specifically for public surveillance. Based on the area of the city, the number of cameras per square kilometer is likely around 155

[a] Data on the number of permanent residents is from the 2016 National Economic and Social Development Statistics Bulletin of each city

[b] Information on the number of cameras in various cities came from the following reports: Shun (2017), Wei (2017), Chen (2017), Dake (2015), Aiming (2015), Shuhua (2017)

making for an average of 1 camera for every 11 people. If you live in London, you may be photographed by surveillance cameras up to 70 times a day.[4]

China's population is twenty times that of the UK and it's forty times larger in area. Excluding Hong Kong, Macao, and Taiwan, China has four municipalities directly under the central government, 27 provincial capitals, 334 prefecture level cities, and 2877 county level cities. There are six cities with populations over 100 million. Based on that and a figure of 20 million cameras, that means 1 camera for every 14 people.

Thinking carefully about this phenomenon, we can see that there is something behind a tendency to voyeurism. Even stronger than that is the need for safety and security.

The modern city is one where everyone is a stranger. It's completely different from traditional rural society. This helps breed a certain amount of psychological insecurity among urban residents.

---

[4] *Evening Standard* (2012).

How do we make a city safe? In 1961, American urbanist Jane Jacobs published her masterpiece *The Death and Life of Great American Cities*. In it, she talks about the idea that doors and windows of buildings should face the street. Houses facing away from the street makes the streets less safe because they lose the protection of their "eyes on the street." She observed that neighbors can distinguish between people that belong on the street from those that they don't recognize—and who might warrant closer supervision. Her concept of eyes on the street advocated for small-scale neighborhoods, developed with a variety of businesses. These small neighborhoods and businesses increase opportunities for people to meet, enhancing the safety of the area.

Our present era is now quite distant from that of Jane Jacobs. Global cities are increasingly going down the path of large-scale urban development. Traditional neighborhoods are disappearing, the mobility of urban populations is increasing, and old connections are being slowly severed. Jane Jacobs' idea of eyes on the street is increasingly untenable in our present age, when the value of every inch of urban real estate is soaring.

Without many people noticing, artificial eyes on the street have become the new solution. Looking at this in a global context, most cities are building or reinforcing this key surveillance infrastructure. Unlike the eyes on the street that Jacobs advocated—decentralized and independent human eyes—these new electronic replacements can be networked into a unified architecture. Rather than a grassroots neighborhood watch, surveillance cameras are a top-down solution. These cameras are being installed in cities by not only government bodies, but also by corporations, families, and individuals. Skynet is not one-dimensional. It's at least three-dimensional, having now penetrated the ecology of the city to the finest level. These cameras are now part of the "micromanagement" of urban space.

2.3 Three-Body: The True Dimensions of Skynet

A month after National Day, I was in Chaozhou in Guangdong, visiting a friend. He lived on the third floor of an ordinary apartment building. He had installed a camera downstairs, outside the entrance corridor. A small screen in his living room provided a live broadcast of goings-on below. He told me that he had done this to watch his motorcycle. If he caught someone approaching the motorcycle, he could shout at them through a speaker on the camera, "*Zou me!*"[5] That would usually lead the stranger to flee.

This is another example of the "eye in the sky." When I walked around Chaozhou, I found other cameras installed for the same reason. There were quite a few. Almost all of them were watching parking spaces, building entrances, or stairwells. They

---

[5] Cantonese for "What do you think you're doing!".

invariably ended up capturing more innocent goings-on, as well. That had caused privacy disputes and there were even cases of neighbors suing each other.[6]

In *The Peaks of Data*, I wrote about the precedent for this:

> Nextdoor is a social networking service for neighborhoods. To connect with other users, you need to live in the same area. The site was established in 2010 and currently covers 29,000 communities in the United States. Where I lived in San Jose, there was a neighborhood called Bel Aire-Hillstone. In January of 2013, someone on the site reported a lost parcel. (Some might not be familiar with the American practice of leaving parcels at the door.) There were soon responses from other people that had experienced packages going missing. Someone suggested a locked parcel deposit box. Someone else suggested that they plant a fake parcel loaded with dog droppings. After some back-and-forth, the specific time and date of the packages' disappearance was determined. It happened at four in the afternoon. Everyone promised to keep a lookout. Within a couple days, someone posted to say that they had seen a Porsche Cayenne in the neighborhood, seemingly casing the area for packages. A detailed description of the car was posted. Eventually, someone posted a picture of the Porsche, confirming that it was the car responsible for the missing packages. Two more days went by and someone posted the license plate. Someone recognized the car and contacted the driver to ask if they were involved. The driver turned himself in to police the next day.

I would call this crowdsourced crime fighting. It's carried out by the public and their eyes—an uncountable number of eyes, capable of capturing everything. Now, Nextdoor is trying to use their platform to wire all personal surveillance cameras together (Fig. 2.2).

Many American companies and homes install their own surveillance cameras. When nobody is around, the cameras are capable of noticing irregularities and sending an alert to a mobile phone. Nextdoor wants everyone to connect their own cameras to their platform. The coverage of a single camera is limited but hundreds linked together can form an effective surveillance network that covers the entire neighborhood. If each family installed a digital video recorder, as well, everything the cameras captured could be stored in the cloud. This service is offered free of

---

[6] Two parties surnamed Gu and Dong were neighbors in Tianhe in Guangzhou. Their apartments both had entrances off the same corridor. Since 2013, they have been involved in five lawsuits. In the first case, Dong installed two cameras outside his apartment after finding his locks tampered with. Gu objected to this on the grounds that their family was being surveilled. They took the case to the Tianhe District Court, arguing breach of privacy. After mediation, Dong removed the cameras. The second case was triggered by Dong re-installing cameras. He placed a camera on the ceiling outside his door. This camera also covered public areas. Dong was not open to negotiation with his neighbor. They went before the Tianhe District Court again and, after mediation, Dong removed the camera. In October of 2014, Dong installed a doorbell camera on the inner door of the house. When the outer door was closed, the camera's field of view was restricted to the area directly in front of the door. But when the outer door was open, the camera would have a view of Gu's entrance. Gu sued again. The Tianhe District Court ruled that the camera did not violate Gu's privacy or portrait rights. An appeal was denied by the lower court. But Gu refused to accept the ruling and took the case to the Guangzhou Intermediate People's Court. In September of 2015, they rejected the appeal, too, upholding the decision of the lower court. Gu once again refused to accept the ruling. The case was taken to the highest court in the province. They agreed to a retrial, disagreeing with the decisions of the two lower courts. They ruled that Dong must stop recording. See: "Home surveillance cameras lead to privacy infringement lawsuits," by Shi Youxing in *Procuratorial Daily*, 2017, November 8th.

Aaron Lung from BelAire-Hillstone on 30 Jan

Porsche SUV driving around the neighborhood at 9:57
If this is the same guy who's stealing packages, he passed by my
house a 9:56 this morning

♥ Shared with only BelAire-Hillstone in <u>Crime/Safety</u>

View or reply · Thank · Private message

 Fabio Angelillis from BelAire-Hillstone
at 11:45 AM

I think it is, since the car has the same rack on the roof (both
toward the back) and it is not a standard feature for Cayenne's
I looked at 200 pics of this type car on the web and not one
hard this configuration...

Thank

Fig. 2.2 Building a smart society and crowdsourced crime fighting

charge, which means that if a household installed a camera, they would have access
to video of the entire neighborhood. The effect of this is that the whole is greater than
the sum of its parts. Connecting the cameras together on the platform also means
that there are more eyes watching the street.

These personal surveillance cameras can be compared to the private economy. The
private economy is always more dynamic than the state sector. This is the second
dimension of Skynet.

There are "eyes in the sky," but there are also "mobile eyes," mounted at various
positions on city buses, taxis, and private cars. This is a mobile Skynet.

Take Hangzhou in Zhejiang and Linfen in Shanxi as examples… In 2014, 8,200
buses in Hangzhou were equipped with four cameras each. That means that there
were 32,800 cameras altogether.[7] The two cameras mounted on the front of the bus
were also capable of monitoring the bus lane for vehicles that shouldn't be there.
Photos of the vehicle could immediately be transmitted to traffic police. According
to reports, within ten days of the December 2016 introduction of the cameras, 230
offenders had been captured in the bus lane by the front-mounted cameras.[8]

The United States has a lower population than China and fewer cars. If the Zhang
Yingying case had happened in China, would it have been so complicated to solve?
Just comparing numbers, it seems simple enough. A friend in the Public Security
Bureau told me about two actual cases he had encountered. They show the power of a
mobile Skynet. One involved a traffic accident in a remote area. The accident resulted
in a death. There was no surveillance at the scene of the accident, but the police put
many officers on the case. Two weeks later, they turned up a dash-mounted camera
video from a passing car and solved the case. In another incident, a car collided with

---

[7] Wang (2014).

[8] Jiang (2016).

an electric scooter. According to surveillance footage of the site of the accident, the car had run the red light. But the car's camera showed that the scooter had also run a light. In the end, both drivers were found to be responsible. If the driver of the car hadn't installed a camera, they would have been found at fault.

Cameras on vehicles have played a key role in determining accident responsibility, but have also been involved in solving other cases. In 2016, a woman at an intersection in Pukou, Nanjing was robbed by someone on a motorcycle. The suspect was clearly an experienced thief. He pinpointed the surveillance camera and kept his face turned away. But the police can be just as clever as the criminals. The Nanjing cop that took over the case was even more experienced than the robber. He reviewed surveillance and noted the vehicles that passed through the area around the time of the crime. He ran their plates and contacted the drivers, asking if they had dash cams. Sure enough, one of the driver's had captured the incident on camera. The suspect was tracked down within a day.[9]

The three-way intersection that the robber had chosen was not a coincidence. He had realized that passing cars would be the best way to hide his appearance. But he had not taken into account the dash cams of those passing cars. He took the cars to be not much more than scenery, but they turned out to be reliable witnesses against him. They were even more reliable than human witnesses, too, since they had the ability to play back the scene just as it had happened. If the criminal had realized that he was being recorded by passing cars, he would likely have never attempted the crime.

These surface cameras are also part of China's Skynet. According to the Ministry of Public Security's Traffic Management Bureau, there are more than 304 million cameras installed on cars as of June, 2017. Private vehicles accounted for 156 million of those. Vehicle cameras cost in the range of 300–400 yuan (around $45–60 U.S. dollars). They will likely become standard equipment on cars in the future.

State Skynet, private Skynet, and the mobile vehicle Skynet are all part of Chinese Skynet's "three-body." If you add up all these cameras, it is in the hundreds of millions.

I believe that we should actively work to create the conditions necessary to strengthen the network of the private Skynet. Networking cameras together makes them even more powerful. But this kind of networking must be different from the model offered by online streaming platforms. The Public Security Bureau should establish a registration system to keep track of cameras, especially in key locations such as corridors, shared exits, and parking lots. If a crime is committed, they will be able to use this database to trace a suspect's path and provide evidence for his prosecution. The police will be able to contact camera owners, as they did in the above cases. If cameras were registered in a database, it would save police the laborious process that's currently required. Neighbors could also share video with each other to help community safety. In the future, car cameras will be networked into the system through the cloud, as well.

---

[9] Yutong (2016).

2.4   What the Internet Can Learn from Skynet

The construction of Skynet began in the 1990s, tracking the development of the Internet. At the beginning, the two collected data in similar ways.

Skynet's initial data collection method was simple, direct, and without permission. Cameras were installed indoors and outdoors and collected video 24 h a day. Anything that their lense captured was stored. Few people noticed the cameras and there were no signs to indicate that they were there. Without any awareness of them, how could the surveillance be deemed consensual? That is why I refer to it as "without permission."

Internet companies also carried out a similar collection without permission. If you use their website or install their mobile app, you are surrendering your data. There is no obligation to inform users of this, let alone ask their permission. As for where this data is stored, who's using it, and what they're using it for—the user has no say over any of this.

The situation with Skynet was that it was recording public spaces in order to maintain public safety. It was simply recording what was happening out in the world, which, in principle, anyone has the right to do. That makes the lack of permission more understandable. On the other hand, collection of data by Internet companies was for commercial reasons. It was data about consumption habits and behaviors. The lack of consent to the data being collected is much less tenable than in the case of Skynet.

It was also not Internet companies that broke with this system first. In April of 2007, the city of Beijing adopted the Measures of Beijing Municipality for Administration of Image Information Systems for Public Security, which stipulated that any collection of surveillance in public places must be indicated with a sign. The city began putting up bilingual signage in place with surveillance cameras, reminding individuals that their behavior was being recorded and to govern themselves accordingly.

After 2007, most cities in China adopted similar practices. Signs saying "This area is protected by video surveillance" became ubiquitous. This was tantamount to seeking permission to record people. Entering into the space was to give approval to surveillance. To some degree, this honored the public's right to privacy and disclosure of surveillance.

Shortly after Google Maps Street View launched in the United Kingdom, they were challenged by a local human rights organization called Privacy International. The group filed a complaint with the British Information Commissioner's Office (ICO), stating that despite attempts to censor the faces of pedestrians in Street View, individuals were identifiable in the more than 200 photos they had collected. Their request was that Street View be banned in the UK, although that was eventually denied. The IOC pointed out that taking pictures of people on the street was not against the law, unless it was done in such a way as to harass them.

There have been similar problems in Japan. Street photographer Masato Seto frequently made pictures on Japanese streets and his images of subways had won him the prestigious Kimura Ihei Award. After the pictures were published, Seto was sued by one of the women in them. He eventually lost the case. To avoid similar awkward situations and to deal with the dark side of the Internet, new rules were made. One of those rules was that any device capable of taking a picture couldn't be operated in silent mode. If you take a picture with your phone, it will make an audible click that cannot be muted. A short time later, South Korea established similar regulations.

The sound effect informs the subject of the photograph that their image has been captured. People still have the right to record images in public, but they must inform the other party. If you use the images for any purpose, you must seek the consent of the subject. This is done because the image might affect a person's reputation, safety, privacy, or livelihood.

Everyone is familiar with the Miranda Warning given by police when they arrest a suspect: "You have the right to remain silent. Anything you say can be used against you in court." Interrogation of a suspect is also a form of data collection. This warning is to inform the suspect that data collection has begun and that they should govern themselves accordingly. The Miranda Warning has become a global norm.

If recording images and sounds follow these rules, then the collection of other types of data should comply, as well. However, Internet companies have ignored global norms. They are unwilling to give up data collection without permission.

On January 3rd, 2018, Alipay placed on users' financial statement a checkbox to agree to their service agreement. One of the terms of this agreement was that users consent to data collection by Sesame Credit, Alibaba's credit scoring program. The agreement also allowed Sesame Credit to transfer this information to third parties. Most people would click the "I agree" box without reading the agreement.

The agreement attached to the annual statement was intended to allow users to expose their accounts and grow the Alipay brand, causing a chorus of angry voices. There was public anger at the practice of hiding privacy agreements. Some people went as far as to call it "data theft" (Fig. 2.3).

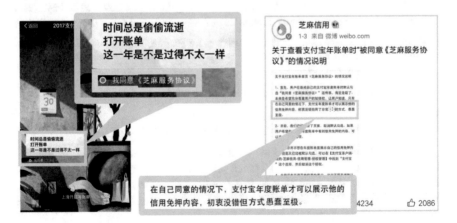

Fig. 2.3 Small print in an Alipay statement

Left: "I agree to the Sesame Credit terms of Service" was checked by default on Alipay's statement (partially enlarged).

Right: The apology issued by the company on January 3rd, 2018.

Alipay immediately apologized for their misjudgement. It was more than a misjudgement, though, since it violated China's Cybersecurity Law, Standard on Personal Information Security, and other laws. The apology was about all they did, since there was no change to the policy. In March of 2018, Alipay was fined 180,000 yuan by the People's Bank of China for "inadequate protection of consumers' data, inadequate disclosure of information gathering, and improper use of personal financial information."

In an interview conducted at the Fortune Global Forum on December 6th, 2017, Tencent CEO Pony Ma revealed something about his company's facial recognition software. He said that although many companies had developed the capability, Tencent was far ahead of them. He revealed that there were more than a billion photographs uploaded to Tencent every day, with that number doubling or tripling on holidays. Most of the photos are selfies. Because people sometimes begin uploading selfies from a young age, Tencent can watch them grow up. That means that Tencent can predict their future appearance. The photos were uploaded for users' own purposes, but Tencent used them to develop facial recognition software.

On New Year's Day of 2018, Geely chairman Li Shufu expressed his own concerns about data and privacy. He said: "Pony Ma has to be reading our conversations on WeChat. He can do as he pleases. This is a major problem." This criticism touched on the very lifeblood of Tencent. They gave an official reply, stating that Tencent "does not save WeChat users' chat histories and they are stored only on the device."

This reply seems to contradict what Pony Ma said. Tencent doesn't save chat histories but does it save pictures uploaded to the platform? If not, then where do those billion pictures come from every day?

On a Tencent website introducing its big data analysis capabilities, it claims to have 800 million user portraits. If they are not saving data, then how could they claim the capability of analyzing data? This is what it says on the official Tencent website:

We offer real time statistical data analysis services for your applications, capable of monitoring performance, status, user behavior and segmentation. Our data visualization assists you in making key decisions.

Based on Tencent's 800 million QQ user portraits, we offer a complete, reliable solution to identifying your customers.

We provide statistical analysis to fully integrate with any business scenario.[10]

In simple terms, today's Internet companies suffer from "multiple personality disorder." They want to show off to potential customers their capabilities to analyze big data, but they also want to hide that from their own users, never revealing the source of their data. When they are confronted on data collection, they try to wriggle out of it, claiming innocence.

---

[10] From Tencent's official website: http://data.qq.com. Site accessed: May, 2018.

Of course, these problems extend beyond Alibaba and Tencent. Toutiao, Sina, JD, and Baidu have all been charged by the Cyberspace Administration with violating rules about data collection and also being unable to justify the collection of this data.

What the Internet needs to learn from Skynet is that the age of data collection without permission and notification has ended. Internet companies must inform users, tell them what data is being collected, and get consent for its use.

2.5 Fog Computing: The Battleground for Artificial Intelligence

The public's perception of Skynet is likely to experience a major upgrade.

As I've already written, the idea is almost universal that mastering artificial intelligence will be the summit of the next technological competition. Everyone realizes that this will bring major social change. In the future, people will realize that surveillance cameras can be used for more than data collection, and the debate over them will extend beyond security and privacy. People will discover that Skynet is an important factor driving the development of AI. Skynet will become a battleground for AI.

Information can be divided into three categories: images, words, and sounds. When it comes to human perception, the senses of sight and hearing are the strongest. The sense of sight takes up 80% of perception, with the remaining 20% given over to sound. Images are more plentiful in our lives than sounds, and we are more sensitive to them. The present goal with artificial intelligence is to replace humans with machines, which first means giving them visual and auditory capabilities that will match our eyes and ears. To be more specific, the ordinary camera must become a smart camera.

I believe that smart cameras, rather than robots, will be the first large-scale application of AI. The smart camera can not only record but also analyze images and sounds. Someday in the future, you might look into a camera and have it call out your name.

Smart cameras are what will give people the sense that they have entered an AI age.

Skynet is the largest source of image data. That makes these "eyes in the sky" a precious resource for developing AI. This AI will allow Skynet to transform from passive data collectors to analysts.

There is another reason why the popularization of smart cameras is inevitable.

Cameras must be networked online. This allows surveillance cameras to become a city's retina. Skynet will become fully integrated, allowing each lens to be linked together, allowing data to be shared. All cameras in a city can be interlaced as a complete system that allows complete management of positioning, tracking, and management. This allows for unified adjustment and capability sharing.

The biggest hurdle to networking cameras is storing and transmitting data. Video files are large. Compared to recording sound, the size of a video file is 100 times larger, and it would be 10,000 times that of a text file. The transmission of video files is a problem that the Internet must face. Sending videos now accounts for about 90%

of network traffic. With millions of cameras running 24 h a day, the amount of data they produce could overwhelm the network.

Digression: Project Dazzling Snow pushes surveillance networking forward

The Chinese government has promoted the networking of surveillance. In May of 2015, the National Development and Reform Commission issued "Several Viewpoints on Strengthening the Constructruction of the Public Security Video Surveillance Network." Their suggestion was that by 2020 all surveillance networks have "full coverage, complete networking capability, nonstop avail- ability, and sophisticated controls." These are goals to ensure public security, optimize transportation networks, serve city management, and innovate social governance. "In key public areas," the report stated, "video surveillance must reach 100% coverage. In fields related to public security, networking must reach 100%." That was the reason for the Central Political and Legal Affairs Commission launching Project Dazzling Snow (also known as Sharp Eyes) in October of 2016.[11] One month later, 80,000 surveillance feeds in Quanzhou, Fujian were connected to the Internet. That number included 30,000 feeds from the Public Security Bureau. In addition, there were 50,000 secondary and tertiary feeds from township-level governments, as well as campus and traffic surveillance. As a result, Quanzhou was selected as one of the government's Public Security Video Surveillance Construction Network Application Demon- stration Cities, the only prefecture-level city in Fujian Province to receive the honor.[12]

In 2011, Flavio Bonomi, former head of advanced R&D at Cisco, introduced the concept of "fog computing."

Fog can be all around us. To extend the metaphor, "fog" is the "cloud" at ground level. Fog computing is pervasive and ubiquitous. Cloud computing centralized data for processing, but fog computing gives devices on the periphery the ability to analyze data. That reduces the need for data transmission and dependency on central servers.

The smart camera relies on this concept. It is not just a lense but also a computer capable of analyzing what it sees. They float in the air like "smart dust." At present, a traffic surveillance camera transmits data back to a centralized server to be analyzed. This consumes network capacity and introduces lag. But if the camera has the ability to analyze data, it can make real time decisions about what it is recording. Once it makes its judgment, it can transmit the data to an on-site police officer or send it

---

[11] The project was born in Linyi, Shandong. Throughout 2013, it was implemented experimentally in Pingyi County, then throughout the entire city by 2015. Coverage extended even to remote rural areas. In 2016, the Central Comprehensive Management Office and other departments implemented their own program to develop networked public security video surveillance applications and improve fully integrated crime prevention and control methods.

[12] Kaiyu (2016).

back to the central server. This reduces the demands on the central server and saves bandwidth.

Given the way the system works, I prefer the name "edge computing" to "fog computing." The trend is now toward empowering the edge of networks. More capability will be deployed to front-end equipment, along with key data. Back-end computing resources can be saved for more detailed integration and analysis.

2.6 Using Pictures to Track Cars: Tracking the Trajectory of Hundreds of Millions of Vehicles

Let's return to the question raised at the beginning of this chapter: what made the Chinese police officers I spoke to so confident that they could track the car in the Zhang Yingying case?

A trajectory is a series of time-stamped locations. To track a moving object, you must know its trajectory. That has been known since ancient times.

Humans discovered that the stars were in motion. Out of curiosity or fear, the ancients cast their gaze heavenward. Generations tracked the position of celestial bodies without the assistance of scientific instruments. They began to trace the trajectories of these objects.

By tracing the trajectory of heavenly bodies, mankind discovered that there are 365 days and four seasons in the solar year. That helped them to make important decisions about planting and harvesting, migration, coping with floods, and holding religious rites. In addition to the skies, mankind turned to the study of local fauna. They tracked these animals' movements to learn the laws that governed their diets, migration, and reproduction. Meteorologists and ecologists studied the trajectories of hurricanes, too, and ocean currents.

Digression: Tracking trajectories, data, and heavenly bodies

The most basic way to chart a trajectory is to record the points through which an object moves. In 1605, Johannes Kepler (1571–1630) was looking down a chart showing the position of the sun and stars over the span of centuries. He had been trying for years to figure out laws of motion based on what he had observed. Whatever he did, the laws he formulated did not seem to match the charts. But one night, he realized that planets move in ellipses, rather than perfect circles. Suddenly, the data recorded by Nicolaus Copernicus (1473–1543), Tycho Brahe (1546–1601), and other astronomers made sense. Kepler was correct. He could finally chart the trajectory of heavenly bodies, which led him to establishing his three laws of planetary motion, laying the groundwork for Newton's discovery of universal gravitation. That is how he earned his nickname: Legislator of the Stars.

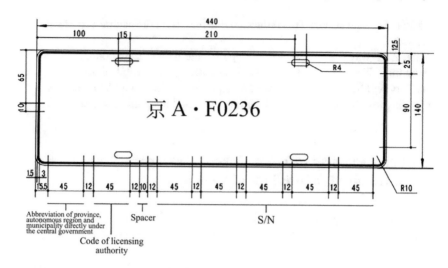

京 A · F0236

Abbreviation of province, autonomous region and municipality directly under the central government

Code of licensing authority

Spacer

S/N

Fig. 2.4 Motor vehicle license plate standard (GA 36-2014)[13]

The movement of pedestrians and vehicles is the most apparent manifestation of urban dynamics. The volume of vehicle traffic is larger than the volume of pedestrian traffic. That is most obvious at intersections. With the help of license plates, it becomes easy to distinguish individual vehicles. If a city has enough cameras at intersections, Skynet can take a series of time stamped photographs. With the vehicle identified, as well as it's driving speed, vehicle appearance, and driver characteristics to help differentiate them, we can track the trajectory of the vehicle, even if it falls outside the coverage of Skynet.

When it comes to tracking vehicles, Skynet relies on nodes. These nodes refer to main intersections. These nodes are not synonymous with smart traffic enforcement. In the case of the latter, photos are only taken from the rear of vehicles that violate traffic laws, like running red lights. But the nodes capture more than 99% of passing traffic. The only omissions are due to vehicles moving too quickly or being obscured by nearby cars. Unless lighting conditions are particularly bad, license plates can be identified in most pictures (Fig. 2.4).

In 2017, various regions began promoting license plates with QR codes stamped in the upper left corner. These QR codes hold the same data as the number plate and are all unique. A QR code can be scanned much quicker than standard optical character recognition. It is an easy way for police to sort out fake plates or unlicensed vehicles.

The work of license plate recognition is similar to the system to differentiate counterfeit bank notes. That is, there is a unified template. China has taken steps over the past two decades to standardize license plates down to the millimeter. In 1992, the banning of letters *I* and *O* was designed to stop confusion with the numbers *1* and *0*. A law introduced in 2008 made it illegal to have a license plate frame without

---

[13] Image source: People's Republic of China Vehicle Number Plate (GA36-2014).

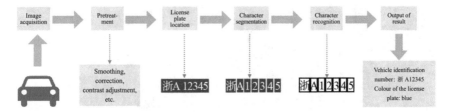

Fig. 2.5   Identification principle for vehicle license plate

at least 5 mm of clearance. License plate frames also cannot contain any Chinese or other characters, decorative patterns, or images. These regulations were to facilitate license plate regulation.

The principle here is the same as the finish line of a race being moved toward the runners. It's about making the final goal easier to reach. AI improves but what it's attempting to do also becomes easier. We are transforming our surroundings to make them more comprehensible to AI (Fig. 2.5).

I and *1* are easy to confuse, as are *Z* and *2*, *S* and *5*, *B* and *8*, and *D* and *O*.

As the technology for reading license plates has improved, companies have introduced software that allows users to upload the image of a car and have its trajectory through Skynet nodes charted. You can see exactly where a car has traveled.

In 2014, Taobao launched Pailitao, which allows users to take a picture of an item and have identical or similar products displayed on the site. Similar technologies have spread across e-commerce platforms, including JD and Amazon, as well as search engines, like Google and Baidu. Vehicles have recognizable features, and so do most items that consumers might train their cameras on. Those differentiating features could include symbols, brand names, or colors and shapes. The principle of identifying cars and products is the same.

Starting in the second half of 2016, the technology for optical recognition of plate numbers began to be widely implemented. The transportation hub at Shanghai Hongqiao, integrating the flow of airport, railway, car, subway, and bus traffic, has used the technology, too. In 2016, they found that at least 70 vehicles were entering and exiting more than 100 times in a month; a single car was found to have entered and exited the hub 516 times. This pattern showed that the cars were not being used for personal trips. When contacted, most of the owners admitted that they were running illegal taxis.[14] In May of 2017, Hongqiao Airport implemented a smart recognition system, so that vehicles coming and going no longer had to take a ticket but would be photographed and identified by AI. Violators could be found immediately (Fig. 2.6).

Left: The item being captured.

Right: Search results matching the item.

---

[14] Qin (2016).

Fig. 2.6 Taobao's Pailitao function, which can identify objects and recommend products. *Left* The item being captured. *Right* Search results matching the item

I don't think it would hurt to expand and flesh this out: if all the nodes and parking surveillance systems in the city were connected, there could be complete coverage of traffic. Having worked in traffic enforcement, I know that cars involved in the commission of crimes are frequently dumped in parking lots for long periods. In the future, this method of evading detection will no longer be feasible. The Public Security Bureau will be able to make use of universal coverage to track the trajectory of a car through the city. Nothing will remain concealed from investigators.

However, linking nodes and parking systems is not simple. Parking lots and parking garages are under different management and ownership. That makes networking them quite difficult. But the trend is clearly towards closer linkages, so this will be accomplished eventually.

Getting a handle on vehicle trajectories also allows for better emergency management. Police can find themselves in pursuit of vehicles involved in crimes or accidents. We've all seen the chase scenes in movies. But with Skynet, police can let their quarry get a bit further away, then plan for a safer point at which to make an interdiction. That will help avoid creating traffic chaos, as well.

Fig. 2.7   A driver is nabbed for using their phone while driving

Other than reading plates, the current system can also recognize characteristics like vehicle model, make, color, and the appearance of the driver. The 2016 General Requirements for Artificial Intelligence Traffic Monitoring and Recording mandates driver photographs no less than 50 × 50 pixels. That has allowed the system to be applied to other things, like identifying drivers that are using a mobile phone. On June 1st, 2017, more than 140 nodes on expressways in Zhejiang began looking for distracted drivers. In the past, due to the difficulty of catching drivers in the act, these laws were hard to enforce. Within five days of starting the initiative, more than 4000 drivers were charged.[15] As you can see from Fig. 2.7, even if the vehicle is going over the speed limit, pictures will still be clear.

A similar situation was faced with the issue of catching drivers not wearing seatbelts. In October of 2016, Zhejiang launched a provincewide initiative to use cameras to catch those not buckling up. Images were captured by high-speed cameras at nodes and the pictures sent directly to traffic police.[16] In June of 2017, 942 cameras in Hangzhou gained this new capability. In less than a month, the system detected 19,000 cases of drivers not wearing seatbelts.[17]

As the technology advances, face recognition technology will also be brought into the equation, allowing the number plate and the driver to both be positively identified. We will be able to know who is driving a car, but we can also take a picture of an individual and match them to vehicles and times they were driving.

That brings us to another question: if Skynet can trace the trajectory of a vehicle, can it trace the trajectory of a person? For the camera, a license plate and a person's

---

[15] Jianguo (2017).

[16] Haodeng (2016).

[17] Chongyuan (2017).

face will be as easy to differentiate, so why couldn't an individual's path through the city be traced?

The ability to track a person as they move through the city has been keenly anticipated by local governments around the world, and especially by their police forces. Although it was formerly impossible, today, it seems like it might be just around the corner.

2.7   The Secret to Why External Hard Drives and Eye Drops Sold Out at the Same Time

"He will kill anyone within ten paces/And will not stop till he has gone a thousand miles./Shaking the dust from his clothes, he goes into hiding,/To shroud in secret his person and his name." So go the lines by the great poet Li Bai. He is describing a man that can take the life of anyone within ten paces. He cannot be tracked on his flight of a thousand miles. He disappears without leaving a trace. We have no idea who he is.

For a long time, it was hard to track someone's movements. Before we entered the Information Age, comprehensive tracing of someone's trajectory was virtually impossible. To track someone, you needed to rely on direct observation, by inter-viewing eyewitnesses, or through a log book or diary. It was difficult and frequently unreliable.

With the advent of mobile phones, the situation has changed. The way our phones work, they have to stay in touch with a cellular station tower. These towers are linked together to form a network. Each tower operates on a particular frequency. As a phone moves, it switches between cell towers.[18] Status updates, text messages, and phone calls are carried onto the network by the tower. The radius of each station's coverage varies from hundreds to thousands of meters. From this, the approximate location of a phone can be determined. But the accuracy is dependent on the radius of coverage. You can tell the general area where a phone is, but not much more.

But if cell tower station location and image searches from traffic surveillance could be combined, we would have a "chemical reaction."

We can see that in a case from Zhangpu, Fujian. In 2010, police in the city began trying to find ways to stop an auto theft gang operating in the area. In January of 2012, they tried using surveillance to track a recently stolen car. They determined that the car had passed a particular intersection at 4:10 in the afternoon before going to Chaozhou, so they pulled records from a nearby cell tower, covering the time when the car would have been in the area. They noticed several calls made from a number in the coverage area to a Chaozhou number. The trajectory of the vehicle

---

[18] As of September 2017, China has approximately 6 million stations, including 4.4 million 3G/4G stations. There are approximately 1.3 billion mobile phone users (142 million 3G users, 947 million 4G users).

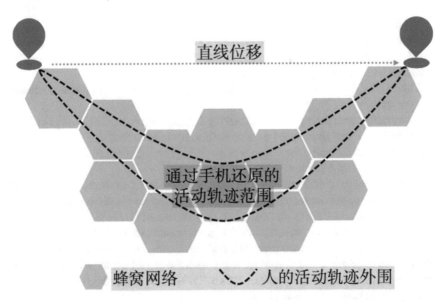

Fig. 2.8   Triangulating from cell towers can give a rough sense of a phone's trajectory

was established. By using data from the cell tower and traffic surveillance, it wasn't long before a suspect was arrested (Fig. 2.8).[19]

> In order to save on building more towers, telecommunications companies try to expand the coverage of cell stations. Coverage areas are roughly hexagonal, which leads to a honeycomb pattern. From the station a phone is connected to, its location can be narrowed down to one of these areas.

Digression: The significance of tracking trajectories goes beyond public security

The applications of this technology go beyond cops and robbers. Studying the movement of people in the urban environment can help us to analyze other phenomena. If we can fully grasp how people are moving through the city, steps can be taken to avoid traffic congestion and optimize transit. Advanced tracking can also tell us about the lives and behaviors of citizens, even down to the identities of the people they associate with. Some experts believe that despite the random or chaotic appearance of individual behaviors, they can be predicted with some confidence. The success rate can rise as high as 93%.[20]

---

[19] Chenghao (2013).

[20] González et al. (2008).

Most smartphones now have built-in Global Positioning System (GPS) capabilities and Wi-Fi, so we can go beyond cell tower triangulation. These methods are accurate enough to locate a phone down to meters. But this is data that can only be collected with the user's knowledge and permission.

This is one of the challenges of tracking phones. Here, the power lies in the hands of the user.

If someone shuts their phone off, tracking becomes difficult. That brings us back to the importance of Skynet. As we said before, it's hard to walk for long in the city without being caught on a surveillance camera. Skynet doesn't need express consent to capture your image, either.

The UK was the first jurisdiction to build a surveillance network, and they were also the first to use facial recognition. This became even more pressing after the July 7th, 2005 London bombings. In that attack, four British Al Qaeda members exploded bombs in a suicide attack on three London Underground trains and a double-decker bus, killing 52, injuring more than 700, and paralyzing the transit network. Police spent hours combing through footage from the hundreds of cameras inside stations and thousands mounted on the street. It took five days to locate the four suspects. It came down to hours of staring at screens.[21]

When hard drives and eye drops were sold out in stores across Nanjing in 2012, nobody would have guessed that a similar effort was taking place, with police collating and then analyzing hours of video footage to solve a crime. This has become something of a legend in police circles in China. When I worked in Nanjing in 2017, I learned some things that few know about the case.

It started with a robbery on Heyan Road on January 6th of that year. The culprit had gunned down and robbed a man that had just gone into a bank to withdraw 200,00 yuan (about $30,000 USD). The details of the case reminded veteran police officers of Zhou Kehua, also known as "God of Death," a wanted criminal, at large for eight years, and known to be involved in at least eleven murders.

The modus operandi was quite similar to previous crimes that Zhou Kehua had committed, so police surmised that he must be at large in Nanjing. They started making a citywide search for Zhou, but also began pulling surveillance footage from across the city, attempting to determine what his trajectory might be.

The Public Security Bureau copied surveillance footage to thousands of hard drives that were distributed to officers. Within a day, Nanjing stores had been stripped of external drives. The police went through the footage frame by frame, scanning faces. When they couldn't take it anymore, they leaned back to put in eye drops. The search would continue until the suspect was taught.

A similar thing had actually happened in Changsha between December of 2009 and June of 2011, when Zhou Kehua was being sought for three crimes he had committed in the city. Kuang Zhengwen, then head of the team poring through footage at the Changsha PSB recalls that they had 1,000 of their members going through hundreds of thousands of gigabytes of surveillance over the space of two months. "It would be the same as every officer watching a hundred movies in a

---

[21] Knight (2005).

week," Kuang Zhengwen recalled. He would retire to his office in the evening and sort through footage that had been flagged. He would study these scenes intently, scanning the screen for anything suspicious. During that time, he would only sleep two or three hours a day. In the morning, he would rush out to any locations identified in the footage, making measurements, and investigating. It took three months before Kuang Zhengwen finally had a clear frontal photo of Zhou Kehua. This was a major break in the case. In 2017, Kuang was named to a list of "China's Top 100 Criminal Investigators."

Nanjing police were able to determine that Zhou Kehua had arrived in the city at least twenty days before the crime and had frequented the area around the bank. In the following days, he had traveled on buses and bought daily necessities at a supermarket. However, just like in Changsha, by the time they began tracking his trajectory, Zhou had already left town.

A similar search was staged in the United States in the wake of the Boston Marathon bombing. On the afternoon of April 15th, 2013, two bombs exploded at the site of the race, killing three people and maiming 183 others. The police immediately began working on footage, trying to figure out who might be responsible. Unlike in the Zhou Kehua case, social media usage had progressed a bit by the Boston Marathon bombing. Thousands of people had been taking pictures at that moment. So, in addition to watching footage from the surveillance network, the police also had to comb through video and images from thousands of phones belonging to people that claimed to have seen the suspect.

Finding clues was down to cops watching screens. A team was set up to watch videos around the clock. In one case, an officer watched the same video 400 times. That effort, with the help of eyewitnesses, helped the police find a suspect three days later. They were able to find a clear picture of him on surveillance video (Fig. 2.9).[22]

Top: Images of two suspects taken from surveillance at the scene.

Bottom left: Front and profile image of the bombing suspect.

Bottom right: The official ID photos of the suspect, which facial recognition software did not turn up.[23]

With a picture of the suspect, shouldn't it be possible to compare it to their database with facial recognition software? Afterwards, the police realized that they had driver's license pictures of both suspects on file, but they didn't turn up in any searches. The facial recognition software was essentially useless.

Without any high-tech recourse, the police turned to old-fashioned methods: press conferences, publishing pictures in newspapers, and offering rewards. In the end, a relative of the two suspects recognized their faces and provided police with their names.

Here, we've seen stories about manual surveillance video searches from China, the UK and the United States. At this time, surveillance networks were simply eyes in the sky, without any intelligence. The situation today is very different. There are far more

---

[22] Montgomery (2013).

[23] Gallagher (2013).

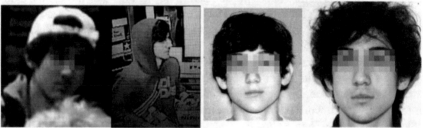

Fig. 2.9 Images of the suspects in the Boston Marathon bombing

cameras, generating even more footage, but they now send their recordings to the cloud, where they can be easily downloaded for analysis. If it happened today, a case like Zhou Kehua's would not require that surveillance be distributed on external hard drives; the footage could be easily pulled from the cloud. The practice of manual searches of footage is also a thing of the past, as facial recognition software has greatly improved. There won't be another run on external hard drives and eye drops.

How would the killer in Li Bai's poem fare? He wouldn't be able to evade capture by fleeing thousands of miles or hiding his identity. The next question to ask is: How useful is face recognition? Will the system ever be capable of identifying a human face as easily as it identifies a license plate? Will we be able to give Skynet a face and have it give us an identity and location?

If we can do that, what changes will we see in social management and business? Is a future of advanced facial recognition what we have to look forward to? What will be the ultimate significance of this technology?

In fact, AI for facial recognition has been going on for six decades. But, dear reader, I leave that story to the next chapter.

References

Zhou Aiming. (2015, October 12). "Nanjing installs 295,000 'Skynet Safety' points." *Nanjing Daily*.

Shen Chen. (2017, May 9). "Guangzhou Public Security Xueliang Project has increased the rate of cases cracked by video surveillance sevenfold." Southcn.

Wu Chongyuan. (2017, June 23). "Still driving without a seatbelt? You could join 20,000 other drivers already nabbed by a new automatic detection system." *Zhejiang Online.*

Quan Chenghao. (2013, December). "Research into vehicle and mobile phone tracking for criminal investigation." *Journal of Hubei Public Security Academy.*

Hu Dake. (2015, August 14). "The secret behind Hangzhou's eyes in the sky: the special aid they've given investigators in cracking cases." Hangzhou.com.

Evening Standard. (2012, April 12). We're watching you: "Britons caught on CCTV 70 times a day."

Gallagher, S. (2013, May 7). "Why facial recognition failed in the Boston bombing manhunt." *Ars Technica.*

González, M.C., Hidalgo, C.A., & Barabási, A. (2008, June 5). "Understanding individual human mobility patterns." *Nature.*

Hunter-Syed, A. (2021, January 7). 45 Billion Cameras by 2022 Fuel Business Opportunities. LDV Capital.

Huang Shun. (2017, May 19). "Surveillance used in 50% of criminal cases." *Shenzhen News.*

Sun Haodeng. (2016, October 18). "Zhejiang expressways implement system capable of automatic detection of drivers not wearing seatbelts." *Hangzhou Online.*

Liu Jiang. (2016, December 20). "By eleventh day of operation, cameras on Linfen buses have captured images of 230 offenders." *Shanxi Evening News.*

Xu Jianguo. (2017, June 6). "Startling new technology catches 4000 traffic offenders." *Qianjiang Evening News.*

Knight, W. (2005, July 13). "CCTV footage shows London suicide bombers." *New Scientist.*

Lin Kaiyu. (2016, November 24). "Quanzhou: Public security departments begin integrating video surveillance networks into Project Dazzling Snow." *Xinhua Online.*

Montgomery, D. (2013, April 20). "Police, citizens and technology factor into Boston bombing probe." *Washington Post.*

Qin Qianqiao. (2012, April 9). "The nation's first video surveillance police squad monitors Wuhan streets: Skynet and terrestrial forces are finally linked." *China Police Daily.*

Wang Qi. (2014, September 21). "More than 8000 buses will be fitted with surveillance equipment before the end of the year." *Daily Business.*

Li Qin. (2016, June 23). "E-vehicles used for 'black taxis' may have licenses lifted." *Shanghai Morning Post.*

Chen Shuhua. (2017, September 24). "'Smart Skynet' helps criminal investigation clearance rate reach new heights." *Quanzhou Evening News.* (All sources originally in Chinese.)

Yang Wei. (2017, August 20). "Facial recognition used to identify 39 suspects." *Yangtze River Daily.*

Liu Yutong. (2016, September 6). "Dash cam used to solve motorcycle robbery of Nanjing woman." *Nanjing Daily Online.*

Chapter 3
The Face and AI

Abstract Our faces have become an entrypoint for artificial intelligence. The portraits you post of yourself on social media will be saved, transmitted, analyzed, and processed countless times in the cloud. Nobody will tell you this or ask for your permission. There are processes that nobody notices. The seeds of AI for facial recognition were planted in Silicon Valley and they blossomed into the world. This chapter will explain how that happened.

> In March, the peach trees flower, and in September the chrysanthemum blossom. Each waits patiently for its time to come.—Chinese proverb

In this chapter, we will travel back through the history of facial recognition, exploring how this tidal wave formed from a modest wake.

Unique features and peculiarities are common in the natural world. Nature never tires of crafting these differences, like in the lines of leaves or in the pattern on a palm. They can be so minute as to not appear to the naked eye, requiring specialized tools. But people can tell faces apart quite easily. More than a hundred billion people have lived on Earth,[1] but no two of them had perfectly identical features.

The face represents ourselves. It is also the way we express ourselves. Subtle changes in facial expression can convey nearly every emotion we might feel. This adds to the mystery and profundity of the human face.

This is why the human face has always fascinated artists. From Ancient Egypt's Sphinx to Da Vinci's *Mona Lisa* to Munch's *The Scream* to modern photography, the face is the eternal theme of artists. The face is also a way for the artist to reflect other things. English artist and critic John Berger (1926–2017) summed it up like this:

> Whatever the painter is looking for, he's looking for its face. All the searching and the losing and the re-finding is about that, isn't it? And 'its face' means what? He's looking for its return gaze and he's looking for its expression—a slight sign of its inner life.[2]

---

[1] American demographer Carl Haub believes that before the adoption of agriculture pushed out hunter-gatherers, that global population would have hovered around five or ten million. According to the censuses that were irregularly carried out in the first century AD, the global population can be estimated at around 300 million. Putting a relatively high birth rate on these figures, we get an estimate of around 106 billion people have lived on Earth. See: Xuefeng (2001).

[2] Berger (2002).

Z. Tu, *The New Civilization Upon Data*,
https://doi.org/10.1007/978-981-19-3081-2_3

I understand the power contained within the human face, too. I have learned some of this from experience presenting and attending lectures. Any set of slides should contain a human face, preferably smiling. There is nothing more enticing or warm than the sight of a human face smiling back at us.

Mankind long ago recognized that the face is the most important, most fundamental, and most convenient way to recognize each other. Naturally, making realistic faces became an early challenge for computer graphics; in the age of AI, recognizing faces has become a key part of developing machine vision (this is also called computer vision). In both applications, development started with photographs. It's interesting to consider what came before the photograph, though: portraits drawn or painted by artists.

3.1   Photographs Are the Key: Building the Identity Society

Painting is also a way to record things. Compared with writing, it can be a more perceptive medium. If we take *Along the River During the Qingming Festival*, *The Night Revels of Han Xizai*, or the *Last Supper* as examples, we can see that artists can record social norms, customs, and daily life, as well as pursuing esthetic perfection. Before we had photography, artists left a record of our life and times.

Truly great painters were rare; having a portrait painted was a great luxury. Whether in Occident or Orient, portraiture was restricted to monarchs, nobles, and military heroes. Seeking the finest and most accurate portrait possible, the Qianlong Emperor (1711–1799) even sought out the Italian Jesuit painter Giuseppe Castiglione (1688–1766). In rare cases, the privilege was afforded to members of other groups, including courtesans and fugitives. There is even a story about a portrait changing the fate of two nations. The subject of that portrait was a woman named Wang Zhaojun. Emperor Yuan of Han was looking for a concubine and had the painter Mao Yanshou sketch the candidates to save himself time on the process. The candidates bribed the painter to hide their defects. But Wang Zhaojun refused. Mao Yanshou deliberately painted her in an unflattering way. She was never visited by the Emperor and her beauty was only discovered when she was selected to be sent to the Xiongnu as a concubine as part of a peace treaty. When Emperor Yuan of Han realized what had happened, he killed Mao Yanshou in a rage.[3]

Emperor Yuan of Han might have missed out on the beautiful concubine. But the Xiongnu appreciated her. They felt as if the Emperor must put great stock in the treaty. Wang Zhaojun understood what was required of her and she fulfilled her mission. After Wang Zhaojun joined the Xiongnu rulers' household, the borderlands were peaceful for decades.

---

[3] This story comes from *Notes of the Capital of Chang'an*, written by Ge Hong in the Jin Dynasty. It's a historical record of goings-on in the capital of the Western Han.

A portrait can change history. The literati of past dynasties said it, too. There are even poems about it, like the matchless wisdom of Wu Wen, who advised that a painting was worth its weight in gold.[4]

As I said, fugitives were also sometimes the subject of portraits. When the rulers of a nation sought a criminal, they would commission a portrait to put up in key locations, so that the officers of the law would know who to watch out for. In *Chronicles of the Eastern Zhou Kingdoms*, Feng Menglong (1574–1646) recorded a case like this. In 522 AD, King Ping of Chu ordered a portrait to be painted of the famous general Wu Zixu. He was trapped in the city. The fear turned his hair white overnight, which made his portrait inaccurate, allowing him to slip out undetected. This is a very popular story in China.

There is a similar plot in *Romance of the Three Kingdoms*. Cao Cao attempted to and failed to assassinate the warlord Dong Zhuo, who immediately ordered that documents and portraits of the fugitive be distributed.[5] It reminds us that this sort of portrait of a fugitive was the birth of facial recognition. The hunt for criminal suspects still relies on this method, although photographs have replaced paintings.

Photographs are what made computer facial recognition possible.

Frenchman Louis Daguerre (1787–1851) invented photography in 1839. He used a copper plate with a photosensitive silver layer to capture and fix images. This laid the foundations for modern photography. He called this "writing with light." It gave birth to a new age, where a moment or a life could be "reborn" in the form of a photograph.

The process caused a sensation. It was unbelievable that a moment could be frozen in time, and even reproduced endlessly. It was nearly unbelievable. A paper in Leipzig wrote: "To try to capture fleeting mirror images is not just an impossible undertaking, as has been established after thorough German investigation; the very wish to do such a thing is blasphemous."[6]

Photography provided mankind with something to rival writing and painting. It was a new frontier in the ways that we could record and preserve moments in history.

The first thing that photographers aimed their lenses at were human faces. Many were eager to have their images preserved for posterity, so there was a fashion for portraiture across Europe and the United States. In 1860, Queen Victoria sat for her first photographic portrait. The image sold more than 500,000 copies across the country.[7] That was only the beginning, though, as the technology steadily improved.

---

[4] See: "Ming Fei" by Wu Wen, a Qing Dynasty poet.

[5] *Chronicles of the Eastern Zhou Kingdoms* and *Romance of the Three Kingdoms* are not quite accurate. The descriptions in them of drawing portraits of wanted men are not quite accurate. Researchers have concluded that this practice was not widespread until the Ming and Qing. The specific phrase to describe it first appeared in a history of the Ming Dynasty. Before that period (1368 to 1644) lack of paper and printing technology limited the circulation of these sorts of images. The invention of commercial paper production by Cai Lun, an Eastern Han official, did not take place until 105 AD, so it's very likely the story of Wu Zixu's portrait was fabricated.

[6] Wenfang (2004).

[7] Ibid.

By 1888, Kodak had invented commercial film, reducing the cost of photography, and opening up portraiture to a new audience, which flocked to newly opened studios.

At this time, the West was reaching the climax of its Second Industrial Revolution. These advanced societies of the West glowed from the bright incandescent bulbs that began to be put up in their cities, and they seemed to roar with the sound of machinery. It was a new age. In the Technological Revolution being undergone, establishing identities became increasingly important. In an agricultural society, people are mostly confined to their homes and the land they cultivate. They rely only on their plot of land. They live and die close to that land. For them, simply having a name is enough to identify them. But an industrial society is much larger. There comes division of labor. A worker is no longer tied to the land, either. A worker's labor, welfare, mobility, safety, and finances all rely on establishing identity. To do that on the basis of a name alone, it would be hard to determine who is who. There is no way a state could manage its population.

Photographs began to be used for personnel files. They became a new management method, providing managers with a way to fix someone's identity. Faces became the key to determining identity. This was a huge advancement. During the First World War, governments of Western states began to compel citizens to submit photographs for their records. Portraits became key to handling various documents. This collection of portraits from citizens became the basis for facial recognition decades later. Even now, the biggest customer for facial recognition technology remains governments. That is true around the world.

Face recognition is identity based on biometrics. But face recognition is not the most mature or most accurate biometrics identification technology. Although collecting and processing of fingerprints was complicated and time consuming, advances in optical scanning technology had improved the process. But in contrast, collecting photographs was much quicker. Portraits began to pile up in archives. Here, we see how the track was paved for facial recognition.

3.1.1 Wandering Out of Silicon Valley and into the Geometric Age

The earliest research into facial recognition technology was done in Silicon Valley. In 1960, artificial intelligence expert Woodrow Wilson Bledsoe (1921–1995) helped found Panoramic Research, Inc. (PRI), funded in part by the Department of Defense (DoD), charged with researching facial recognition. At that time, the funding for artificial intelligence was coming mostly from the DoD and intelligence agencies.

The problem set for PRI was very clear: give a photograph of an individual to a computer and have it find all matches in a database, making the job of manual comparison quicker.

The reason why the DoD and intelligence agencies wanted the capability was to help them in keeping faces attached to identities. If someone committed a crime, then changed their name and traveled to a remote jurisdiction to apply for a new identity, new documents could be issued to them—the same face but a different name.

Bledsoe proposed using geometric features of human faces to differentiate them. This laid the foundation for the subsequent two decades of research.

The geometric method was to measure the position and size of various features on the face. The size, location, distance, and ratios could be used to recognize the face. This requires a lot of manual assistance. The first step is to mark the pupils, corners of the eyes, hairline,[8] and other key points on a face. After that, various calculations were made about the width of the mouth, the distance between edges of the eyes, the distance between pupils, etc. These ratios and distances were used to compare faces.

Bledsoe was later selected as the chairman of the American Association for Artificial Intelligence (AAAI). He kept pushing research forward but recognized that facial recognition was a major challenge: a person's face in different photographs might be at different angles, completely changing the geometric measurements recorded from each one. His method was to analyze silhouettes. It's a principle similar to silhouettes in Chinese folk art. Bledsoe drew the silhouette as a contour line, took a point as reference, and compared it to the same point on another image. Compared to viewing the face from the front, the silhouette recognition system was rather crude, producing a lower rate of recognition. There could be additional factors to consider, including light intensity, angles, facial expression, and changes with age.

In 1969, three Japanese scientists announced that they had created a program capable of determining whether a picture contained a face. In other words, they could pick a face out of a complex image. In 1970, a Stanford University researcher took things another step forward: he was able to extract the face from pictures. In 1973, Takeo Kanade (1945–) announced that he had identified 15 faces in 40 photos of 20 people.[9] This was the best result achieved at the time. Takeo Kanade became a key figure in the Geometric Age (Fig. 3.1).

> When confronted by two photos of the same person, the computer could determine if they were the same person by identifying key points.

Generally speaking, this early facial recognition, rather than a process of artificial intelligence differentiating faces, was computer-aided manual facial recognition. Research into facial recognition showed it to be a very difficult proposition. Although each face is usually composed of the same features, the combination of these features can be almost infinitely varied. Just as no two snowflakes are the same, neither are two human faces. The difference between human faces and the diversity is amazing.

Even Solomon Shereshevsky (1886–1958), the greatest mnemonist of the twentieth century, struggled with this task. He could memorize extremely complex mathematical formulas, huge matrices, and even long portions of texts in foreign languages. In one experiment, he was able to recite a list of 70 items from memory. If he heard something once, he could repeat it, front to back and then back to front. He had to admit that he couldn't remember faces:

---

[8] More specifically, this refers not to the hairline but the part of the hairline that is the lowest on the face.

[9] Kanade et al. (1973).

Typical sequence of analysis steps:

a.   Top of the head

b.   Cheeks and sides of the face

c.   Nose, mouth and chin

d.   Chin contour

e.   Face-to-face line

f.   Nose lines

g.   Eyes

h.   Face axis

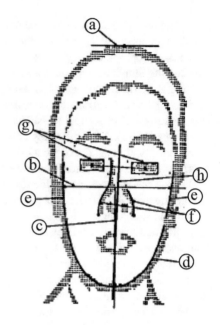

Fig. 3.1 Eight geometric features facial reognition relied upon in the 1970s[10]

> They're so changeable. A person's expression depends on his mood and on the circumstances under which you happen to meet him. People's faces are constantly changing; it's the different shades of expression that confuse me and make it so hard to remember faces.[11]

A face is hard to remember because its characteristics are hard to accurately quantify. I asked Wang Feng, a Chinese mnemonist, famous for recalling 300 digits from memory, for his opinion on this. He told me that numbers and words are easy to memorize because of their composition: a number is nothing more than a combination of digits from 0 to 9, and words are limited to a particular set of characters. These are easily encoded by our brains. The human face is more of a challenge. There are nearly infinite possibilities in the arrangement of features. It is very difficult for us to encode a face into our memory. We might recall that someone has a large face, wide set, large eyes, double-folded eyelids, thick lips, etc. but these are qualitative rather than quantitative statements. That leads to challenges recalling and recognizing faces.

From this, you can see what the Geometric Age was all about: the machine was capable of quantifying the human face in such a way as to give us the data necessary to recognize them. The calculations required were beyond the capabilities of the human mind.

By the late 1980s, new methods began to appear, but facial geometry was still the basis. That meant that the process still began by finding the position of the eyes, determining the area of the face, converting the image into grayscale (color does not

---

[10] Ibid.

[11] Luria (1987).

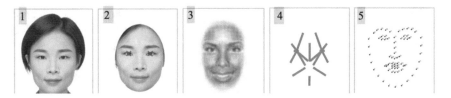

Fig. 3.2 The logic process of facial recognition

come into the calculations), then extracting various features to compare to images in a database.

Computer scientists in the 1960s had already figured out that the key to facial recognition was isolating features. If computers were capable of doing it themselves, could we solve facial recognition? Researchers began moving from the method to the tools, working on developing AI capable of doing just this (Fig. 3.2).

> Steps in facial recognition: 1. Get a photo. 2. Determine position of eyes, then crop and convert to grayscale. 3. Convert image into a form that can be placed in a template that the software can recognize. 4. Compare template to images in the database. 5. Output results.

3.2   Can Machines Learn?: The Struggle for Artificial Intelligence

So, if we say that photography solved the problem of recording and facial recognition makes use of these recordings, then artificial intelligence is the way that we can make automation of facial recognition an automatic process.

Artificial intelligence as a field of research emerged from Dartmouth Summer Research Project on Artificial Intelligence, a 1956 summer workshop at Dartmouth College, organized by John McCarthy (1927–2011). At the time, McCarthy was still a year shy of his thirtieth birthday and only an assistant professor. His insistence on the title for the emerging field—artificial intelligence—is the reason we still use that name today.

As the name suggests, artificial intelligence is the synthesis of human intelligence with computers. The fact that it implies machine intelligence might replace human intelligence has led to disputes.

When the field was conceptualized, there was the belief that machines could be given logic, reasoning abilities, and a degree of intelligence that would allow them to assist or replace humans in making certain judgements. For that reason, mathematical logic was the focus of early research, with the idea that this artificial intelligence could solve theorems. However, it was later deemed that this was not sufficient. Edward Albert Feigenbaum (1936–) changed the focus of determining whether or not a machine was intelligent from formal reasoning to an emphasis on knowledge. From that point on, with the support of researchers taking up his viewpoint, AI entered a new age focus on both logical reasoning and knowledge. Feigenbaum won the ACM Turing Award in 1994 and is often called the "father of expert systems."

However, mankind soon discovered new paradoxes in artificial intelligence. Expert systems focus on specialized knowledge, drawn from a professional field, but human knowledge is boundless and constantly updated. Computer scientists must summarize knowledge and teach it to computers. This is a never ending process. Soon, a new question arose: Can we have a machine that learns by itself? If so, how would that be done?

There were 20 participants in the Dartmouth Summer Research Project workshop, including IBM's Arthur Lee Samuel (1901–1990). In 1952, IBM had released the IBM 701, the first commercial scientific computer and its first series production mainframe computer. A short time after its release, Samuel developed the first computer checkers program. The program was proof that computers could not only process data but also show a certain degree of intelligence. IBM's stock rose 15% on the program's release.

Samuel continuously improved his checkers AI. He programmed a search tree of board positions that could be reached from the current state, but he also wrote programming that would allow the program to optimize its choice based on memorization of positions it had already seen. Samuel believes that machines would one day use a learning process that would allow them to achieve an intelligence similar to human intelligence. In 1959, he proposed the concept of machine learning.

Samuel believed that machines were capable of learning, and his checkers program was proof. But skeptics pointed out that checkers is a relatively simple challenge and limited in scope. Mathematicians proved that as long as both players played a perfect game that the result would be a draw. The problems that mankind hoped AI could tackle were more complex than checkers. It was difficult to replicate the limited success of the program (Fig. 3.3).

Fig. 3.3 Samuel plays checkers against an IBM computer

The checkers program Samuel created defeated many human players. Compared to chess and Go, the defeat of human champions by a computer came very early. The Chinook program developed by the University of Alberta in 1989 defeated human champion Marion Tinsley (1927–1995) in 1994. In 1996, the first human chess champion was defeated by AI, and the turning point came in 2016 for a Go champion.

The debate over whether or not machines can learn has divided researchers into two camps.

The first school is behind what could be called Conventional AI, attempting to imitate in a machine the intelligence of humans. This school believes that intelligence is unique to the human mind, so AI can only be achieved by mimicking it. To develop intelligence in a machine requires giving computers thought processes similar to our own.

The other school is behind Computational Intelligence. They believe that machine intelligence does not necessarily need to mimic human intelligence. For example, birds and airplanes both have wings, but they operate differently, and one is much faster than the other. The idea is to develop the computational systems of existing computers, using math and logic to construct rules for them, which might allow them to eventually learn.

Of course, as with any other debate, there is some middle ground. The centrists in the AI debate tend to support Conventional AI in theory, but they hold that within a short period of time, machine intelligence will develop past the cognitive mechanisms of human intelligence. Without going beyond Conventional AI, development of machine intelligence will stagnate, so it's a pragmatic choice to keep an eye on development of Computational Intelligence.

McCarthy, the founder of the field, is a firm believer in Computational Intelligence. He said that AI relied on certain rules and logic. In the invitation letter to the Dartmouth workshop, he stated the purpose of the gathering:

> The study is to proceed on the basis of the conjecture that every aspect of learning or any other feature of intelligence can in principle be so precisely described that a machine can be made to simulate it.[12]

Implied here is the idea that all knowledge must first be described. The only knowledge that can be transmitted to machines must first be encoded in human language. Knowledge beyond that cannot be simulated by machines.

McCarthy is completely uninterested in imitating the structure and mechanism of the human brain. He doesn't want anything to do with psychology, cognition, or neurology. He has publicly stated that the study of AI should be separate from the study of human behavior. AI should be a branch of computer science, rather than psychology.

From the beginning, Computational AI dominated the mainstream, but there was a great deal of diversity of opinion, even with individual researchers. Some of the debates in AI have gone beyond the field to enter philosophy. Marvin Minsky (1927–2006) was another organizer of the Dartmouth workshop, and he came to be regarded as one of the founders of the field. He believed that the brain could be completely simulated; the

---

[12] McCarthy et al. (1955)

brain was nothing more than a machine. "I bet the human brain is a kludge," he said. "Humans are nothing but meat machines that carry a computer in their head."[13]

From this, we can see that Minsky belonged to the school that believed AI must mimic human intelligence, but he was also the one that dealt a serious blow to Conventional AI. In 1969, he helped plunge AI into a troubled decade.

3.3   "I Don't Know What It Means, but it Sounds Amazing"[14]: The Rise of Deep Learning

Conventional AI came from the study of the human brain and nervous system. At the beginning of the twentieth century, mankind discovered that the elementary particle of neural function is the neuron.

The human brain has hundreds of billions of neurons, each with its own function. They are connected to each other and process information as a network. The neuron has a portion that receives information, called the dendrite. Each neuron has multiple dendrites but there is only one channel for transmitting information, called the axon. At the tail end of the axon are many terminals, which are connected to the dendrites of other neurons to form a synapse. This means that one neuron can receive transmissions from multiple neurons, process them, and transmit the information on to other neurons.

So, there can be multiple input signals but only one output signal. Any given signal is determined by information received from other neurons. The study of this neurological structure provided inspiration to AI researchers. They tried to imitate it in machines as a way to give them the ability to make decisions (Fig. 3.4).

In 1957, Cornell University's Frank Rosenblatt (1928–1971) proposed the concept of perceptron. The algorithm received data from multiple sources, then used it to generate new information and transmit it on.

Later, researchers in neurology discovered that the strength of connections between neurons can vary in strength. The difference in synapse strength impacted how information was transmitted: some information had a large impact and some had a small impact. As a result, computer scientists introduced a new concept: weight, referring to the strength of a connection between two nodes. The output of a neuron can be expressed with a mathematical formula:

$$Z = G(a_1 w_1 + a_2 w_2 + a_3 w_3)$$

Here, w represents the weight of each input and G represents an unknown sum function. When countless perceptrons are connected together, they function like a human brain. It was the first time that a computational model was used to imitate

---

[13] Kolata (1982).

[14] This term, usually shortened further, is Internet slang, usually used somewhat ironically, reacting to particularly esoteric arguments.

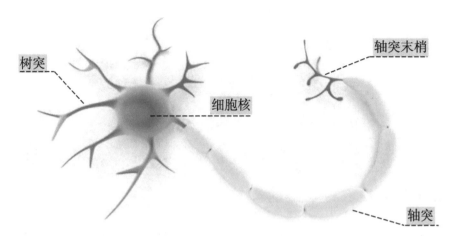

Fig. 3.4 The structure of neurons

the information processing method of the brain's synapses. These were called neural networks at the time, but they can be more correctly understood as the ancestor of the later artificial neural network model.

Over the next several years, neural networks became a key focus of mathematicians, and a controversial method of machine learning. Prior to neural networks, AI was developed by programmers writing code, telling a computer what to do. These were basically If–Then statements. For example, the command might be: if X is greater than 100, then go to the 25th line of code and execute a particular mathematical operation. Or, if a certain condition is met, then output a certain parameter. This is setting rules for the computer. It's an example of top down thinking.

Machine learning does the opposite. It clarifies the independent variables—the input—and dependent variables—the output. At first, there are no rules set, but after being fed data, the machine can begin to infer possible and optimum rules. These rules are about the weight each independent variable is given in the functional relationship. Once this functional relationship (another way to understand this is as a set model) is determined, it can be used to forecast the future. This is bottom up thinking.

Machine learning is the autonomous capability of adjusting the weight given to variables and confirming the model or functional relationship being used. The other goal is to draw down on the point where the output matches real data. In simpler terms, machine learning involves a computer digesting a large amount of data into a self-organizing and self-learning mathematical model that can then be used to predict the outcome of other scenarios.

This process is very close to the human decision-making process. As we've already seen, data is a record of objective reality. This record contains cause, appearance, and effect, representing past experience. With machine learning, these experiences are put into a model that can be used to analyze future events (Fig. 3.5).

In this way, the computer is not obviously smarter than their creators, but more capable. People can only consider four or five independent variables at the same time

Fig. 3.5 Human logic and machine learning

before they are overwhelmed. But computers can weigh hundreds or even tens of thousands of variables simultaneously, performing huge and complex calculations very quickly. That is hard for the human brain to match.

A complete neural network could have hundreds and thousands of perceptrons. Each perceptron must make its own calculations, so a large amount of computer power was required to run a neural network. But through the 1960s and 1970s, as these networks were proposed, computer power was limited. During this time, Minsky published a monograph titled Perceptron, speculating on the future of the neural network. Minsky noted that a single-layer neural network was only capable of simple linear classification and even exclusive disjunction calculations (also called exclusive or, a logical operation that is true if and only if its arguments differ) were beyond its abilities. If you increased the number of layers, even by only a single layer, the amount of calculations necessary would increase exponentially. At that time, this seemed an impossible task. There seemed to be no future for multilayer neural networks (Figs. 3.6 and 3.7).

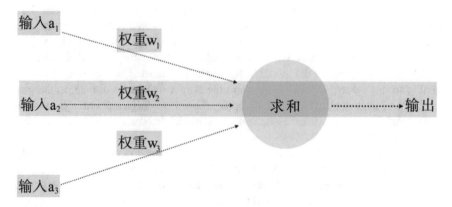

Fig. 3.6 The structure of the perceptron

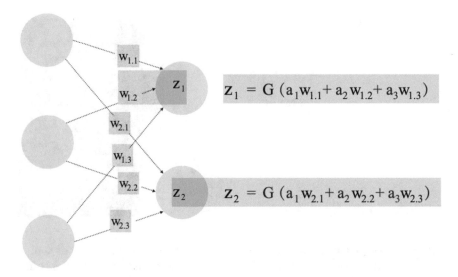

Fig. 3.7 Single layer neural network

At the time, Minsky did not expect that computer power would rapidly increase tens of thousands of times. His judgment turned out to be wrong, but he was restricted by computer power of the time. As a giant in the field, his word carried weight. His pessimistic view of neural networks became common wisdom. Many researchers and laboratories gave up on neural networks.

The next decade is often called the "AI winter." Through the 1970s, if a paper included "neutral network" in its title or abstract, most journals and conferences would reject it. Neural networks were out in the cold.

There were, however, a few researchers that persisted through this AI winter, and the most notable is Carnegie-Mellon's Geoffery Hinton (1947–).

Hinton came from a background in psychology and got obsessed with cognitive science. For decades he had been focused on neural networks, even though everyone around him, including his PhD advisor suggested he abandon the pursuit and confine himself to the field of mathematical logic.

From early in his academic career, Hinton had been interested in the idea that the brain does not store memories of objects or concepts in a single location, but distributes them across a range of neurons. When recalling a concept, the brain does not rely on a single neuron. Concepts and neurons have a many-to-many relationship: a concept can be defined by multiple neurons, and a neuron can participate in the recall of multiple concepts.[15]

What he called distributed representation became a core concept in the development of neural networks. Here's an example: when we hear and need to recall the meaning of something like Changbai Mountain (the name means, literally, "long, white" mountain), it's a process that requires multiple neurons. One neuron gives

---

[15] Hernandez (2014).

Fig. 3.8 Geoffery Hinton[16]

us the first half, meaning long. Another neuron supplies the color. A third neuron supplies the category of the object. When the three neurons are activated at the same time, we can reproduce and understand the concept of Changbai Mountain.

In 1986, Hinton and American psychologist David Everett Rumelhart (1942–2011) proposed the backpropagation algorithm (BP) to train multilayer neural networks, which greatly reduced the number of calculations required. This was a solution to Minsky's criticisms and helped revive neural network research in academia.

The multilayer network idea made possible by BP was very powerful. BP allowed the system to continuously analyze an almost infinite number of complex functions. This was made possible by the fact that computer power had increased exponentially since the 1960s. Even though they had been locked out in the AI winter, neural networks returned (Fig. 3.8).

> In 1994, Hinton lost his first wife to ovarian cancer. He remarried in 1997 and three years later, his second wife was diagnosed with pancreatic cancer. He was convinced that AI would revolutionize medicine. He expects that in the future, anyone will be able to request their genome sequenced for around a hundred US dollars (the current cost is ten times that). He also speculated that radiology would soon be taken over by AI, as well.

The high water mark came in 2006. That year, Hinton published a paper in *Science*, pushing forward the concept of deep learning.[17]

The depth here refers to the number of perceptrons. Hinton believes that as the number of layers increases, network parameters would increase. More parameters would mean that the network had stronger simulation capabilities, allowing it to approximate the nonlinear process of human thought. Hinton also proposed changes to the way AI systems were trained by adding a pre-training process. In this process, a point in parameter space can be found that allows the machine to much more quickly determine optimum weights. From there, the results can be fine-tuned until

---

[16] Shute (2017).

[17] Hinton and Salakhutdinov (2006).

parameters across the network are further optimized. This allows the number of calculations to be decreased, as well as the time taken to complete them.

Deep learning is the name Hinton chose, indicating that it was quite different from previous neural networks.

Deep learning is an attempt to fully synthesize the mechanism of the human neural network. Each neuron can store information and make calculations. This distributes the functions across the network. Neurons in each layer receive input from the previous layer. After the neuron processes a piece of information, it is passed to other neurons. The strength or weakness of the links between neurons is what the neural network confirms through a learning process. Increasing the number of layers can increase parameters and forge more complex relationships between neurons. Multi-layer networks have multiple nested functions. These networks can create sophisticated models, analyzing complex real world phenomena, and approaching intelligent human responses (Fig. 3.9).

The structure of deep learning is parallel and distributed, self-adapting, and self-organizing. It is precisely this self-adaptation and self-organization that has led to criticism. A neural network can only give what it believes to be the most correct result, but it cannot explain why a certain set of parameters or a certain functional relationship is better than another. Deep learning algorithms are like the brain in this way: we might arrive at an answer without knowing quite how we got there. Deep learning is a black box. Even the designer of the algorithm cannot say how it really works. "I don't know what it means, but it sounds amazing," as the saying goes.

Controversy begets controversy. The amazing but inexplicable feats of deep learning soon revolutionized image recognition. In 2012, Hinton led a team in the annual ImageNet visual object recognition software contest, ImageNet Large Scale Visual Recognition Challenge (ILSVRC). His team used deep learning. Prior to this, most competitors in the ILSVRC used support-vector machines (SVMs). In 2010, the error rate of the top team was 28%, and that fell to 25.7% in 2011. In October

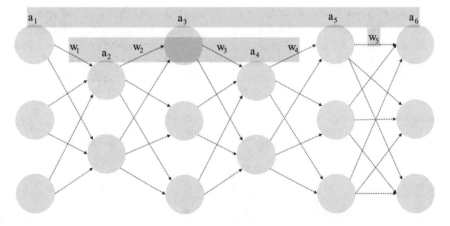

Fig. 3.9 The logical structure of deep learning (multilayer neural network)

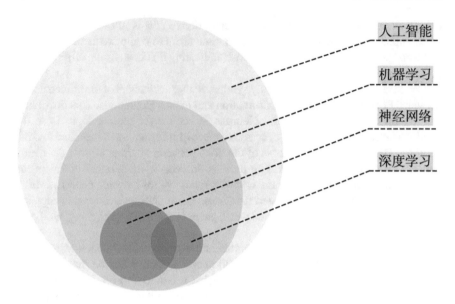

人工智能

机器学习

神经网络

深度学习

Fig. 3.10   The relationship between AI, machine learning, deep learning, and neural networks

of 2012, Hinton's team took that error rate to 15.3%. The second place Japanese team had an error rate of 26.2%. It was a startling moment in the history of artificial intelligence (Fig. 3.10).

After that, miracles came one after another. In the 2013 ILSVRC, the error rate went down to 11.5%, and it was brought to 7.4% in 2014. That number has continuously fallen: 3.57% in 2015, 2.99% in 2016, and 2.25% in 2017, making it lower than the error rate of the average human. In other words, deep learning algorithms surpassed human competitors in image recognition. It was the beginning of a new era.

As we discussed already, facial recognition in the Geometric Age had relied on extracting facial features and comparing them to vast databases that could include millions of photographs. The computer power required was immense. But with deep learning, the problem could be solved. The face could be divided into multiple layers, multiple features, abstract points, lines, etc. and it could compare parts or the whole. Each element would be continuously superimposed and verified, improving accuracy.

Digression: ImageNet Challenge

In 2009, in order to promote the development of the field of machine vision, Fei-Fei Li of Stanford, Li Kai of Princeton, and other Chinese researchers initiated the creation of the massive ImageNet database containing 3.2 million images. It took 80,000 English-language nouns from a language database and assigned

500–1000 images to each. Since 2010, they have held the ILSVRC. The basic rules are the same each year, with participants using 1.2 million pictures from the database (these pictures belong to more than 1000 different categories and have been manually labeled) as training samples for their algorithm, which is then tested with 50,000 new pictures. The picture is marked with the five most likely categories. If the correct answer isn't included in those five guesses, it is marked as an error. The lower the error rate, the higher accuracy mark the software is given. In 2017, ImageNet split the ILSVRC into three categories: object localization, object detection, and object detection from video. By comparing the results from ILSVRC competitors over the years, we can find that with the same number of parameters, deep learning has a higher accuracy rate than shallow networks. Until 2013, the depth of these multilayered networks was still in the single digits. In 2015, the top team used a network with a depth of 152 layers. In 2016, the Chinese SenseTime team won with a neural network boasting 1207 layers.

For example, Google FaceNet,[18] introduced in 2015, can extract 128 facial features, including things like distance between the eyes, length of the nose, ear length, eyebrow top contour line, eyebrow lower contour line, eye width, etc. Google trained a deep learning algorithm and built a multilayer neural network. At first, they gave it three pictures, two of them showing the same person. The algorithm took its 128 features from the three images and began to compare them. This training is done to nudge the system in the direction of linking two of the pictures; even though the points measured will be slightly different between pictures of the same individual, they should differ slightly more than between a picture of two unrelated people.

This step is repeated hundreds of thousands or millions of times, until the neural network can reliably generate the 128 points recorded for each subject. For different photos of the same person, it should give roughly the same function value. When these values approach close enough, the network can determine that they show the same person. For different people, the function value will be reliably different, too. As users, how this is done is not important. What we care about is whether or not the AI can find the same person in different pictures.

Deep learning has greatly improved the accuracy of facial recognition. It has also promoted the use of deep learning facial recognition in commercial applications. In November of 2015, Facebook launched a face recognition function that could identify users' friends in uploaded photos. If the user enabled the Photo Magic function, Facebook algorithms analyzed their recent photos, searching for faces it recognized. It would then prompt the user to share the photos. The accuracy rate reached 98%.[19] In September of 2017, Apple phones started coming equipped with

---

[18] Schroff et al. (2015).

[19] Taigman et al. (2014).

facial recognition tools. Users could use their face to unlock their phone or authorize a payment. According to Apple, they reached an accuracy rate above 99%.

Of course, this is not all down to Hinton's advances. In addition to the continuous optimization of deep learning algorithms, there are two additional factors: the first is the exponential increase in computing power since the 1960s, and the second is the arrival of big data. What big data provides for AI is a large amount of training data. Deep learning allows AI to learn independently, constantly adjusting parameters and functional relationships. Without the rise in computer power and big data, deep learning would never have taken off.

Many believe that deep learning is the most significant breakthrough in AI over the past thirty years. It has reignited the dream of true artificial intelligence. What is imagined could come of AI has expanded. The physiological structures of the human brain have not changed much over the past several millennia, but big data keeps growing and computing power increases exponentially. AI is evolving faster than humans are, which has led to the widespread belief that the former will surpass, rather than simply match the latter.

When we talk about surpassing the human brain, my opinion is that it will take place in the realm of being able to perform calculations faster and being able to analyze information. This expands human brain power, in my opinion, rather than surpassing or replacing it. Human beings cannot create things that are more intelligent than ourselves. This should be the basis for understanding AI.

3.4   Data Track and Field: How the Government Promotes Corporate Innovation

Since 2012 and breakthroughs in deep learning, facial recognition has achieved great successes and its application has expanded. However, facial recognition moved out of laboratories and into commercial applications as early as the 1990s. This built on even earlier successes.

The driving force was the DoD.

Since the Geometric Age, researchers have been working hard on facial recognition, even without the benefit of deep learning. In addition to technical bottlenecks, there were two difficulties: one was the amount of data required to train the algorithms, and the second was that the industry lacked standards to evaluate the benefits or drawbacks of various algorithms.

The United States government resolved both problems. In 1992, the DoD Defense Advanced Research Projects Agency (DARPA), Counterdrug Technology Development Program Office (CTDPO), and Army Research Laboratory (ARL) jointly launched the Face Recognition Technology (FERET) program. The program, which lasted for three years, was meant to leverage government power to push for the establishment of standards in the facial recognition industry. Standardization was the first step to allow the technology to mature, and to eventually bring it to market.

FERET's primary contribution was to establish a large-scale database of images for facial recognition. This allowed researchers from different institutions to develop and test software on a common dataset. That made it possible to easily compare the advantages and disadvantages of various algorithms.

In order to maintain the consistency of the database, the images had to be collected in a controlled way. Each time the subjects were photographed, the same conditions had to be maintained. From August 1993 to July 1996, fifteen phases of image collection were carried out. The database contained 1199 individuals, 1564 sets of pictures, and 14,126 images. The collection of the images took three years because it was necessary to take multiple pictures of the same subject over time to study the effects of age on the face.

Let's say a country wants to develop its track and field talents… The first thing it has to do is build venues for athletes to train and compete. That is the significance of FERET. Its database provided AI players with a place to train. To this day, FERET remains the largest image database of its kind in the United States. It is run now by the National Institute of Standards and Technology (NIST) and still open to use by external applicants.

Between August 1994 and March 1998, DoD organized universities and tech firms to conduct three rounds of evaluations using their database. The results seemed to prove that the technology had matured to the point where it could be deployed in the field. Even when it involved manual intervention, there were still three key problems:

1.   Lighting. It is difficult for AI to match the same face under different lighting conditions.
2.   Time. Even if the program is successful at first, its accuracy decreases.
3.   Geometry. The same as it was since the Bledsoe era, when the face is rotated more than fifteen degrees, accuracy drops.

The FERET evaluation was significant. It promoted the exchange of knowledge in the field, it pushed the players toward a consensus on certain things, and helped create standards. The various players involved in facial recognition also moved toward commercialization. In 1996, Alex Pentland (1952–) of MIT sold his algorithm and established Viisage Technology. Three other MIT researchers started Visionics. Another MIT professor started Miros. These three companies all became standard bearers in the American facial recognition field. Their earliest customers were DoD, as well as government agencies in charge of policing, justice, transportation, and drug enforcement.

The first to officially use facial recognition was the Department of Motor Vehicles (DMV). In the United States, each state has an independent DMV that issues driver's licenses in their own particular format. This led to a problem with one person having multiple licenses in different states. If someone was no longer eligible for a license in one state, they could travel to another jurisdiction and get the necessary documents. Some of those people even worked as commercial drivers. There was also the issue of social security fraud, with people using multiple identities to claim benefits they were not entitled to. Fugitives could also get driver's licenses issued by the DMV in other states as a way to hide their identity.

In 1997, New Mexico and West Virginia undertook a project to standardize and digitize all of their driver's license photos so that they could quickly be scanned by DMV computers. They also began networking DMV management systems across states to allow interjurisdictional confirmation of identities.

By 2015, most states had facial recognition software and were impressed with the results. With the launch of New York's program in 2010, 14,500 drivers were discovered to be holding two or more licenses. These were mostly fraudulent, and they resulted in investigations of 9500 people and the arrest of 3500. Since 2011, the New Jersey DMV began referring an average of 600 cases a year to prosecutors. There was another benefit: by comparing photos, many DMVs found errors in their data. New Jersey's DMV found 12,500 errors in their database between 2011 and 2015.[20]

In addition to rooting out fraudulent identities and errors, facial recognition also simplified the process of driver's license renewal. Government identity documents have to be replaced every few years and that usually requires going to a photo studio to take pictures, but facial recognition technology allowed the process to be streamlined, with applicants uploading pictures taken with their phones. This is the essence of the "single trip" and "no interview approval" programs being undertaken now in Zhejiang, Jiangsu, and Guangdong.

In order to improve the accuracy of the algorithm, photos should be taken from the front, without any head covering, wearing a neutral expression, and with appropriate lighting. A smile makes it harder for a computer to measure the face, since it changes the geometry. This is another example of mankind adapting themselves to a new age of AI. We must accommodate the machines.

After state DMVs, facial recognition software went into use by drug enforcement agencies, passport management, social security and welfare departments, as well as credit management agencies. For federal police bureaus, being able to send a picture of a fugitive to the immigration bureaus was a major step forward. If a fugitive had gotten travel documents under another name, it would be elementary to confirm that. Facial recognition technology showed great potential.

A commercial market emerged for facial recognition around the year 2000. Facial recognition's momentum didn't last long, however. Another blast of cold air like the one that had frozen AI before was about to blow. This time, the cause was Skynet.

3.5 Historical Vicissitudes: How 9/11 Saved Facial Recognition

There have always been two driving forces behind the development of facial technology. The first is technical, driven by researchers in AI. They need machine vision to create smarter AI. Facial recognition was one of the peaks that the field

---

[20] Bergal (2015).

had to surmount. Another driving force was administrative, coming from government departments that were running surveillance networks. Skynet generates a large amount of data each day. Investigations like those carried out by the police services tracking Zhou Kehua were not a sustainable proposition.

When looking at applications of facial recognition, there are broadly two categories. The first is static identification. We have an example of this from Nanjing's Xuanwu District: in March of 2017, they launched a program that would allow people to authenticate their identity by uploading a photograph through an official WeChat account, along with their name, ID, number, and address. After successful verification using facial recognition, they receive a message allowing them to check into a hotel or access other services requiring ID.

The second application is dynamic identification. This is most often used by Skynet. A face appears in a surveillance video and its identity must be confirmed against a database. This presents two difficulties. The first is that the face might not be captured at the right angle, the ideal light, or with enough sharpness, making matching difficult. The second is that the identity may or may not exist in the database. We don't know for sure, so a threshold must be set for what gets sent for manual comparison. If you set the threshold too low, you end up with too many results, and the system releases a flood of vaguely similar results. But if the threshold is set too high, it's possible that matches will be overlooked.

As I've already mentioned, the UK was the first country to build a surveillance network. In 1998, the first group of companies born from FERET began to enter the British market. London Borough of Newham chose Visionics' program. They became the first region to integrate AI facial recognition into a surveillance network.

Prior to this, public order had been lacking in Newham. Visonics claimed that its system could automatically scan the faces of anyone passing under a camera, compare them against a criminal database, and issue a warning. Newham became a publicity victory for Visionics. They installed facial recognition systems in more than 250 locations and spared no effort to promote their "magic bullet."[21]

The results exceeded all expectations. In the year after the system was installed, the crime rate in Newham went down by 40%. In the following year, it was reduced by another 34%.[22] This is impressive data, of course, and you might think it proves the worth of such a system… But the truth is, facial recognition did not lead to any arrests made by local police. Not a single one! Critics charged that the system was "high-tech snake oil" and next to useless.[23]

I believe that the deterrent effect of facial recognition and surveillance networks could be greater than technical capabilities. When technologies like this are popularized and become well known, they become a social technology. That means, their success is not only down to technological capabilities, but also their influence on mass psychology. The good guys and the bad guys are both afraid of ubiquitous

---

[21] "Visionics' FaceIt software installed in London as part of U.K.'s advanced crime reduction program." (2000, February 20). *Security Sales and Integration.*

[22] "Birmingham city centre CCTV installs Visionics' FaceIt." (2008, June 2). *Business Wire.*

[23] Krause (2002).

public data collection, so the role of surveillance networks are exaggerated. When every moment could be recorded, it tends to raise morality. The more awareness there is of surveillance networks and facial recognition, the greater the deterrence effect they have. This is similar to the chilling effect.

In 2001, Tampa, Florida installed a dynamic facial recognition from Visionics, centering it on a nightlife district with lax public order. Locals protested. Opponents of the program charged that it treated everyone as a criminal suspect.

However, as happened in Newham, the Tampa system did not lead to any criminals being arrested. On the contrary, it led to innocent people being harassed. Due to errors made by the system, the system spit out fourteen alerts in one day, leading to a single citizen being contacted by police ten times. Catching bad guys aside, the system could be used to help find missing children or elderly people, but that didn't happen, either.

In the end, Tampa suspended the program. In their statement, they made it clear that the reason for the suspension was not privacy concerns but the poor performance of facial recognition.[24]

The Visionics' prestige faded. In fact, the Tampa statement was a major blow to all commercial facial recognition companies. The results of implementing dynamic facial recognition in the UK and the United States were a resounding failure. Another nadir was on the horizon.

No one realized that a major historical turning point was quietly approaching. Facial recognition might have lost prestige, but this event would trigger national discussion and an upsurge in application of the technology.

This turning point was September 11th, 2001.

That morning, nineteen terrorists hijacked four airliners.

Two of them crashed into the towers of the World Trade Center; one crashed into the Pentagon; and the final flight crashed in Pennsylvania. These attacks caused 3040 deaths. It was the worst terrorist attack in the history of the United States. The country was plunged into sadness, resentment, and reflection. It was a turning point for many aspects of American politics.

A few days after the incident, two photos began circulating in the media and online. The first was from surveillance footage taken at Portland International Jetport in Maine, showing hijackers Mohammed Atta and Abdulaziz al-Omari going through security. After that came a picture full from surveillance at Dulles International in Washington, showing two more hijackers undergoing screening and boarding their flight. Days before the incident, their names had been added to watch lists, but they still passed all security checks (Fig. 3.11).

Left: Atta and al-Omari at Portland International Jetport. They flew to Boston and hijacked American Airlines flight 11.

Right: The second group of hijackers at Dulles.

What about facial recognition? Why was nobody using it? These questions were raised many times in the aftermath of the terrorist attack.

---

[24] Bonsor, K. "How Facial Recognition Systems Work." Howstuffworks.

Fig. 3.11   Images taken from airport surveillance of the 9/11 hijackers

Two months later, during Senate hearings on technology and terrorism, Senator Dianne Feinstein demanded to know how this had happened:

> How could a large group of coordinated terrorists operate for more than a year in the United States without being detected and then get on four different airliners in a single morning without being stopped? The answer to this question is that we could not identify them. We did not know they were here. Only if we can identify terrorists planning attacks on the United States do we have a chance of stopping them. … For example, in the case of at least two of the hijackers, authorities had pictures of them as suspects prior to the attack and airport cameras actually photographed them but because these cameras did not use facial biometric systems, security was not alerted and the hijackers remained free to carry out their bloody plans.

Dynamic facial recognition, on the verge of being abandoned as a failure, was resurrected after 9/11. Suddenly, there were no voices being raised against privacy violations. Using facial recognition and personal data was deemed acceptable, as long as it was fighting terrorism. It was what the public wanted.

The public wanted it, so facial recognition made a comeback. Visionics immediately released a fresh white paper. They said that facial recognition should be a key part of airport security, but that would require a large amount of data and photos from suspected terrorists. With the right database, facial recognition would work. With static and dynamic facial recognition, terrorists would be caught.

Visionics' white paper avoided an important question: Was their facial recognition technology actually effective? Historical experience shows that these questions usually take at least a decade to answer. At that point, though, it was fifteen years before deep learning would mature to the point that it could aid in facial recognition.

3.6   A Society with Anonymity: The Prospect of Dynamic Facial Recognition

In June of 2017, China Central Television reported that ten intersections in Suqian, Jiangsu had implemented a Pedestrian Red Light Facial Recognition system. If a pedestrian crossed against the light, their photo and some personal information would be displayed in real time on a video screen at the intersection. There would be a delay

Fig. 3.12 Screenshot of the Suqian's pedestrian red light facial recognition system

Fig. 3.13 Crossing signal violators caught in Shenzhen

of about ten seconds. The information would also go to traffic enforcement police, who would confirm details in the household registration database. The report went on to say that within a month of the program being implemented 580 pedestrians had been caught running the light. Accuracy was around 90%. In March of 2018, Shenzhen launched a similar program on a larger scale (Figs. 3.12 and 3.13).

These are examples of dynamic facial recognition in action, and they seem to be working… So, why did the systems in Newham and Tampa ten years earlier fail? This touches on two key issues in dynamic facial recognition. As I mentioned earlier, the accuracy of static facial recognition has reached 99%, which shows its maturity as a technology. But dynamic facial recognition falls far short of static identification. It's even hard to determine the criteria with which to evaluate the technology. The problems lie in the fact that the environment around a subject cannot be controlled, neither can angles or light, and camera quality has not been standardized. The quality of images pulled from surveillance footage for image recognition can range wildly in quality. But facial recognition works best when dealing with a frontal photo, with no headcover, and a neutral expression.

When looking at the crosswalk red light cameras, there are certain aspects that make dynamic facial recognition more successful: pedestrians generally walk in a straight line to cross a street, there is generally good lighting, and there are no obstructions. The camera can be set at the best location to capture faces passing by. The red light camera operates in an environment that is not static, but somewhat controlled. It's a semi-controlled dynamic environment.

Even in a semi-controlled dynamic environment, facial recognition accuracy is not close to approaching perfect. Individual companies claim to have achieved 99% accuracy in static recognition, but with very large databases, 99 and 99.5% represent a large difference. The population of a city can be in the millions, and Shenzhen is already in the tens of millions. That means photos are being compared against large databases, which presents a challenge. It's because of these limitations that local police forces cannot use the output of their facial recognition systems to levy fines. That is why they use the method of public shaming of violators.

This method can be effective, too, of course. According to statistics, traffic accidents caused by pedestrians accounted for 16% of the total, and accidents caused by non motorized vehicles accounted for 33.4%. As I write this book, nearly a hundred jurisdictions in China are experimenting with projects to cut down on pedestrian violations of crossing signals. This is not only to cut down on accidents, but also to improve public safety culture.

At the end of 2013, police in San Diego, California started being equipped with portable facial recognition equipment. These tablets could be used to take a photo of people they made contact with, then confirm their identity against a database. Since 2017, Chinese Public Security officers have also begun using mobile equipment with facial recognition capabilities. The idea behind it is described as, "collection of information is recording information, recording information is analyzing information, analyzing information is discovering evidence." Police can use a photo and compare it to provincial databases, national fugitive databases, and persons of interest. It only takes seconds. At present, this mobile equipment is being popularized by front-line officers in county-level police forces. In the past year or so, many fugitives have been discovered by mobile patrols with this equipment.

In February of 2018, Zhengzhou Railway Police were among the first to use facial recognition glasses. News reports claimed that these glasses could detect criminals among station passengers through facial recognition (Fig. 3.14).[25]

---

[25] "Zhengzhou East: The first police force to invest in facial recognition glasses." (2018, February 5). CNS Photo.

Fig. 3.14 American and Chinese police forces are now using mobile facial recognition equipment

Left: Police in San Diego carry equipment that can quickly determine the identity of people they come into contact with.[26]

Right: Facial recognition glasses in use by Zhengzhou Railway Police.

This mobile equipment provides novel scenarios to apply dynamic facial recognition. In the future, dynamic facial recognition will be taking place at street level, with the police confirming the identities of citizens they're making contact with, and also, when facial recognition is integrated into surveillance, high above, in the monitoring centers of these networks. When the cloud can link together all of these cameras and facial recognition, we can quickly confirm the identity of anyone that they capture, and we can also restore their trajectory through the city.

My own prediction is that dynamic facial recognition will one day become ubiquitous. We will enter an age when everyone can be isolated, recorded, and analyzed at will. When we look back at the popularization of photography in the first decade of the twentieth century, governments generally needed to seek cooperation and permission to take photographs of a person. More importantly, capturing this data was a one-time process. Things are very different now. The government can carry out large scale surveillance, capturing pictures of a person in public, without their consent or even knowledge. No government previously possessed this ability.

This also suggests that mankind is about to lose their right to anonymity in public. In a democratic society, the right of anonymity is related to or forms the basis of many other rights. For example, when the police can train facial recognition equipment on a demonstration or march, their rights to assembly may be infringed.

References

Bergal, J. (2015, July 15). "States Use Facial Recognition Technology to Address License Fraud." *Governing*.

Berger, J. (2002). *The Shape of a Pocket*. Bloomsbury Publishing.

Hernandez, D. (2014, January 16). "Meet the Man Google Hired to Make AI a Reality." *Wired*.

---

[26] Winston (2013).

Hinton, G.E. & Salakhutdinov, R.R. (2006, July 28). "Reducing the Dimensionality of Data with Neural Networks." *Science.*

Kanade, T., et al. (1973, December). "Picture processing system using a computer complex." *Computer Graphics and Image Processing.*

Kolata, G. (1982, September 24). "How Can Computers Get Common Sense?" *Science.*

Krause, M. (2002, January 14). "Is face recognition just high-tech snake oil?" *Enter Stage Right.*

Luria, A. R. (1987). *The Mind of a Mnemonist: A Little Book about a Vast Memory.* Harvard University Press.

McCarthy, J., et al. (1955, August 31). "A Proposal for the Dartmouth Summer Research Project on Artificial Intelligence."

Schroff, F., Kalenichenko, D., & Philbin, J. (2015, June). "FaceNet: A unified embedding for face recognition and clustering." Presented to 2015 Institute of Electrical and Electronics Engineers Conference on Computer Vision and Pattern Recognition.

Shute, J. (2017, August 26). "The 'Godfather of AI' on making machines clever and whether robots really will learn to kill us all?" Photograph by Keith Penner.

Taigman, Y., Ming, Y., Ranzato, M., & Wolf, L. (2014). "DeepFace: Closing the Gap to Human-Level Performance in Face Verification." Presented to 2014 Institute of Electrical and Electronics Engineers Conference on Computer Vision and Pattern Recognition.

Li Wenfang. (2004). *History of World Photography (1825–2002).* Heilongjiang People's Press.

Winston, A. (2013, November 7). "Facial recognition, once a battlefield tool, lands in San Diego County." *Reveal.*

He Xuefeng. (2001, June 8). "How many people have lived on Earth? Estimates put it over 100 billion." *Southern Weekend.*

Chapter 4
High Definition Society: Granular Governance and a Population in Cages

Abstract Societies change. In a data-driven society, we all become like tadpoles, trailing our trajectory behind us. In the cloud, we all have a super file. It records nearly all of your behaviors and is updated each day to stay fully integrated. The super file can be checked or verified by anyone, and it can be fully analyzed by a computer. This super file and its analysis and tracking will work differently depending on whether we are talking about a democratic or authoritarian society. However, that's not to say that democratic societies can avoid abuse. I have already covered a number of these potential abuses but this chapter will focus on one of the positives: with the spotlight of data to shine, crime can be averted. We might finally live in—to borrow a famous phrase—a world without thieves.

A wise ruler has no subjects in his nation that trust things to luck. If the common people are waiting for luck to deliver them, it will bring misfortune to the nation.— Adapted from the *Commentary of Zuo*, "Sixtieth Year of Duke Xuan."[1]

As I mentioned in the previous chapter, governments at present can analyze the path of a car traveling through a city with image and facial recognition, and that ability will only improve and become more widespread.

History proceeds down many paths. In addition to artificial intelligence, another development is being harnessed to increase the capabilities of modern governments. That development is progressing faster than even artificial intelligence.

I am referring to the networking and integration of big data.

Anyone living in an urban environment generates a large amount of data about themselves, but there are artificial divisions in the storage of this data. Data about the same person, the same event, or the same period of time might be scattered across the systems of the various units collecting it. Many companies are now pursuing data integration as a way to better understand consumer behavior and drive profits; governments are now throwing their administrative power and expertise behind similar efforts. Coupled with Skynet's ability to track people in an urban environment, the integration of large amounts of data represents a breakthrough in social governance.

---

[1] This is adapted from a text originally in Classical Chinese. A more literal translation would be: "Virtuous persons above, the nation has no citizens believing in good fortune. A lucky person, an unlucky state."

© China Translation & Publishing House 2022
Z. Tu, *The New Civilization Upon Data*,
https://doi.org/10.1007/978-981-19-3081-2_4

Even in countries like India and China, which both have more than a billion citizens, hundreds of millions of vehicles, and tens of millions of companies, governments will be able to exercise control through the cloud, directing people, traffic, and business. As granular governance gets down to an atomic level, I propose a new name to describe it: single particle governance.

Particularly in a highly centralized society, this model presents serious risks. If used carelessly, it can begin a process of putting people in virtual cages. As more and more data is integrated into the whole and every action is monitored, the bars of the cage become thicker and the gaps between them thinner.

4.1 Pattern Recognition: Entering the High Definition Society

At dawn, I addressed a petition to the emperor,

That evening, I was exiled to Chaozhou, eight thousand li away.

I only wished to warn his majesty of evil things,

Knowing that I might sacrifice my good years.

Watching clouds over the Qinling Mountains. Where is my home?

Snow chokes the pass. My horse won't go on.

You must have had your reasons for coming so far,

So please bury my bones by the riverside.

This is a poem by Han Yu (768–824), written in his tragic final years. The emperor had become a devotee of Buddhism and planned a ceremony to receive a Buddha's finger bone relic. Han Yu wrote a letter of petition opposing it. The response came within hours. The emperor sent him into exile in distant Changzhou. Rather than being given time to prepare, he was ordered to leave immediately.

On January 14th of 819, Han Yu headed south into exile. He was 52 years old. Han Yu had a glorious career before that and had achieved many things. Centuries later, when the Song Dynasty poet Su Shi (1037–1101) inscribed a stone stele for a shrine to Han Yu, he wrote: "His writing reversed the literary decline of eight dynasties." The only way to get to Chaozhou was horseback. He struggled to get through snowy passes in the Qinling mountains. His twelve year old daughter died on the way there.

How long did the journey take? There are very few historical records of the trip into exile, but Han Yu is said to have arrived in Chaozhou on April 25th, a hundred days after leaving the court. Throughout history, many other people faced similar situations. Su Shi, who memorialized Han Yu, was sent into an even more unpleasant and distant exile. He died of illness at the age of 64 in Changzhou in 1101. He had recently been pardoned and was on the way home.

Reading these records, we can't help but marvel sadly at the insignificance of our species. We long for eternity, but we leave almost no trace behind. In ancient times, footprints on a dusty road would have been blown away; and it's even harder now: I recall a singer from the 1990s that had a line about how asphalt was too hard to have any trace of our passing be left behind. Rabindrinath Tagore (1861–1941) wrote: "I leave no trace of wings in the air, but I am glad I have had my flight." With those lines, he made a wish for his generation and every one that came before it.

Nowadays, we often take public transportation instead of walking. If you take a trip, you buy a ticket. At the end of the trip, some people hold onto the ticket stub. This is proof that we have taken a trip. Compared to the past, we have a record of our comings and goings.

Some people collect their tickets. When they are old, they can look back on all the trips they took. Those tickets can bring back old memories. But if they are separated from an individual's remembrance, they have no particular value. They don't tell us the path a person took, but only one part of their journey.

Around 2010, China began experimenting with real name registration for train tickets. Once the reform of the ticketing system was complete, it became a requirement for traveling. All trips made after New Year's Day 2012 required real name registration.

A ticket registered to a real name becomes a piece of data stored in a railway company cloud database. Crucially, an individual's trips can be linked together under a single account through a identification card number; this collected data can be pulled at a future date out of a massive collection of data. Thinking back to Han Yu, it's hard to believe: his long voyage on horseback left almost no trace, but our frictionless journey by high-speed rail leaves countless footprints in a database.

In our unbelievably complex modern society, the value and significance of a journey by public transportation being traceable to a single identity cannot be overstated. One of the key conclusions of this book is that a society's level of civilizational development is directly linked to its ability to record and analyze data (Fig. 4.1).

Controlling smoking on trains has always been an issue. In the days before high-speed rail, when the rail network was dominated by the rudimentary "green skin" carriages, the space between cars served as the smoking area. Although there was some distance between passenger seating and this area, second-hand smoke could still drift through the cars, upsetting non-smokers. On the fully-sealed carriages meant for high-speed rail, smoking had to be banned. A new list of regulations introduced in 2014 banned smoking on electric multiple-unit trains (EMUs) and set a fine of 500–2,000 yuan. That was not a sufficient deterrent for hardcore puffers. Occasionally people would light up and simply request to pay the fine.

It was not until 2016 that the railways found an effective enforcement method: from August 15th of that year, smokers could be hit with a ban. Buying a ticket meant acknowledging an agreement not to smoke. If a passenger was caught twice, they could be deprived of their right to ride.

The ban was very effective. It required an electronic ticketing system and real name registration. That allowed for the offense to be linked to an identity. It became much easier to levy fines.

This sort of social management could be taken beyond smoking bans to social credit management. In 2013, the Chinese Supreme Court began to establish a list of "Persons Subject to Enforcement for Trust-Breaking." According to a February

Fig. 4.1 Tickets from multiple time periods and locations. Upper left: A ticket for the first voyage of the Titanic, which sank in 1912.[2] Upper right: An old style train ticket. It does not have the passenger's name or identification number. Lower left: A modern high-speed rail ticket registered to a real identity. This information is stored in a cloud database

2017 report, more than 6.73 million people had been added to the list, 6.15 million had been restricted from buying plane tickets, and 2.2 million banned from buying train tickets.[3] This is a positive effect of granular social governance.

This has been extended even to taxis.

Since their advent, the world has dealt with taxi drivers committing crimes, especially against female passengers. In 1982, taxi driver Lam Kor-wan was arrested in Hong Kong for murdering four of his passengers.[4]

A woman boards a car and then disappears into the vast city. If anything happens to her, it can be hard to determine true events, unless we have records.

In 2016, DiDi launched a function that would allow passengers to share their travel itinerary. Information could be shared in real time with relatives and friends via WeChat, QQ, or text message. That information included departure point, destination, time of boarding, time of departure, distance traveled, estimated time of arrival, license plate, and real time location. Users could set their DiDi app to automatically share the information to an emergency contact. This function was quite popular.

---

[2] "Titanic launch ticket sells for $35,000." (2012, April 16). *Times of Malta*.

[3] Lin (2017).

[4] A pair of thriller movies—*The Devil Butcher* (1991) and *Doctor Lamb* (1992)—are based on the crimes committed by Lam.

金属货币   纸币   信用卡   电子货币

Fig. 4.2 The development of currency toward data

According to data from April of 2017, it was used an average of more than 200,000 times a day.[5]

In May and then again in August of 2018, the DiDi platform was rocked by incidents of young women being raped and murdered. DiDi announced that it would end its carpool service. Apart from strengthening background checks on drivers, my suggestion was that DiDi should implement a one-touch emergency feature. When triggered, it would immediately send location, passenger information, license plate, and details of the car's appearance to police.

Of course, the data recorded now is not limited to travel itineraries but also behaviors related to payment and purchases.

In Han Yu's time, currency was made of metal. That was replaced by paper currency, which is now making way for credit and debit cards. Nowadays, it is more common to use our phones to pay. Currency is just another data point. We add and subtract numbers on a screen (Fig. 4.2).

As I have already discussed, big data has promoted smart business. Payment behavior is disappearing, along with cash. For example, if we use DiDi to hail a ride, there is no need to actually carry out a payment because the fare will automatically be deducted. The only record of the transaction is in the cloud.

In the past, there were transactions but limited records. Now, transactions are becoming even more simplified, so we must keep records of them.

Payment behavior can be recorded but so can social behavior. In 2016, when Kim Kardashian was robbed in Paris, she was left tied up in a bathroom, while $11 million worth of jewelry was taken. Security experts suggested that one reason she was targeted was because of her habit of sharing her lavish lifestyle on social media. On the day of the incident, shortly after arriving in Paris, she had posted about fashion shows, meals, and luxury purchases. These social media posts revealed her location and what thieves could expect to find.

Imagine for a moment what it would be like if Han Yu had Alipay, Weibo, and WeChat. We would know every purchase he made and each stop along the way. A trail of data would follow him. There would be pictures of him in the WeChat Moments of people he met along the way. There would be a clear picture of his voyage from Xi'an to Chaozhou that could be analyzed by future generations.

---

[5] "DiDi's five major safety measures show results: 30,000 applications rejected every day." (2017, April 7). China News Net.

In the final analysis, Weibo, WeChat, and Twitter are all records. They are individualized, granular, and traceable.

What we wear, what we eat, the places we stay, the ways we travel, and where we go—these are all recorded on the Internet and information platforms. Stored with this data are named, ID numbers, phone numbers, and other individual identifiers, such as bank account number, WeChat ID, height, weight, face shape, etc. Data can be integrated into a complete picture. In the database, an individual has a unique "data presence." This presence can be defined, supported, and endorsed by countless pieces of data in much the same way as sound waves or fingerprints. Each piece of data has its own characteristics. This is what I mean by pattern recognition for data. They can simply be called "data patterns."

Just as everyone's face, fingerprints, heartbeat, blood pressure, and other physiological features and functions are different, every person's social behavior will be unique. This data is continuously collected and analyzed. They can also be integrated. A person's physical condition can be connected to their consumption habits, their social relationships, their credit, medical treatment, and education. This forms an even more unique and complex data pattern. With these data patterns, individuals can be defined and differentiated.

In China, our earliest understanding of pattern recognition was when looking at stone and jade. These materials all have different grains, shapes, and colors. Upon closer observation, they can be differentiated from each other. Today, data generated by people, groups, and objects provide social patterns to analyze. We can see them clearly, twisting together or extending separately; they might be multi-faceted or simplistic. For the first time in human history, these patterns can be extracted and analyzed.

Whether in East or West, human societies of the past were not viewable in high definition. It was always somewhat hazy. Even if there were social patterns, they were too coarse to read.

The problem of a low definition society

I called Pittsburgh, Pennsylvania home for six years. The city has four distinct seasons, home prices are low, and it's quite liveable. However much it has going for it, the city has not devoted much of its resources to attracting tourists. There are not many sites to draw visitors. I faced this problem when friends visited me and I had to find places to take them. My default choice was Amish Country.

The Amish are known for their strict observance of the traditions of their agricultural ancestors. In their communities, the achievements of industrial civilization are pointedly avoided. There are no electric lights, telephones, TVs, or cars. They also refuse conscription into the military, voting in elections, and receiving state benefits. They live a simple, primitive life, holding themselves aloof from the people around them.

I can't help but think of Account of the Peach Blossom Spring by Tao Yuanming (365–427) and its description of a utopia protected from the outside world. When friends visited and they saw people living without electricity or cars, they were amazed. The land that the Amish live on is vast and fertile; old and young live together;

and the atmosphere is peaceful. The Amish reject modern industrial civilization's advantages, but also its disadvantages. Perhaps they can provide some kind of model for us.

If a first impression of the Amish was correct, and they're unique simply because they spurn electricity and the internal combustion engine, this might be correct. But it also wouldn't take much to reform that kind of society into a modern one. It was actually a movie that made me realize precisely how primitive and backwards the Amish are—and it has nothing to do with technology.

The Oscar-winning 1985 American film Witness is about an 8 year old Amish boy named Samuel Lapp. While attending a funeral in Philadelphia, the boy is a witness to the slaying of an undercover police officer. While peering through a crack in the stall, he catches a glimpse of the killer's face.

This is another story about facial recognition.

A detective named John Booker is in charge of investigating the murder. He's a tough man that drags the Amish boy around the city, hoping he'll spot the murderer among the people he sees. The boy keeps shaking his head no. They can't find the murderer. One day, out of the blue, he sees the face he's been looking for: Lapp is sitting in the police station when he spots a framed news report about an officer. He's sure it was the man that killed the undercover cop. The man that Lapp fingers used to work with Book and was previously investigated for embezzlement and selling precursor chemicals that the police had seized. Book guessed that the murder of the undercover cop had been an attempt to cover up those crimes. He immediately went to his superiors to report the case. But it turns out Book's commanding officer is in cahoots with the murderer, so he sets out to kill Book. Shot and wounded, Book takes refuge in Lapp's home in Lancaster County—Amish Country.

Book's commanding officer knows the name of the key witness and that he's Amish, so he mobilizes Lancaster police to track them down. Lancaster police inform the city cop that there's no way they're going to find the boy quickly: there are 40,000 Amish in the county, a third of them share the same surname, and none of them have telephones. They have to go door-to-door, which is not a simple procedure.

In modern times, if we want to find someone, we start with their name, then a phone number. After that, we might find an address. Nowadays, we might try to track down a WeChat account. There can be identical names, but likely not identical names attached to the same phone number or address. But in a society without telephones and a handful of shared surnames, it's hard to tell people apart. This is what I mean by a low definition society.

The governance of a low definition society is necessarily inefficient and uncertain. A low definition society can provide asylum to people seeking refuge, but it can also shelter the wicked. In Witness, Amish Country provides a hiding place for Book. That gives him the time he needs to recover from his injuries. When his commanding officer finally tracks him down, he is ready to fight. The battle between good and evil can begin.

Names were humanity's first attempt to reduce ambiguity and move toward a higher definition society.

Digression: Surnames are the creation of the state

Before the 14th century, most of the world did without surnames. In the Meiji Restoration in Japan, the government found the management of household registrations and taxation unwieldy, since not everyone had a surname. They had to give citizens a time limit to adopt them. As a result, many people now bear surnames related to the area where they lived, their profession, or even vegetables. The Japanese population was not large, but that drive to give everyone a surname resulted in it having the greatest variety in names, with the estimate being as many as 110,000. In 1808, we have another example, with Napoleon commanding French Jews to adopt surnames. In 1849, the Spanish arrived in the Philippines and found that most people there shared a small number of names. It caused a great deal of confusion to them. The first thing they did upon establishing control over the islands was to require every Filipino to register a set, unique title for themselves. In Turkey, this happened much later. It wasn't until the 1920s that Turkish authorities commanded everyone to use the name of their home town as their surname.[6]

Using locations and professions as surnames is very common in human history. Chinese surnames like Qi, Lu, Qin, and Jin, derive from the names of ancient states. Less common two-character surnames like Dong'guo, Nanmen, and Nan'gong come from plane names and directions. English-language surnames like Smith and Baker are from professions.

The reason why states created surnames to differentiate people. Clarity is a prerequisite for management and control. Ambiguity is the enemy not only of science but also the greatest enemy of management. In social management, individuals need to be attached to information in a system of national records. At that point, household registration, taxation, conscription, public security, medical care, welfare, property, proprietary rights, and inheritance can be administered. Clarity is the foundation of state administration.

China completed the process of differentiating people based on names quite early. Terms like "old hundred names," referring to common people, is a sign of that long history. In fact, China was fairly stable in feudal times; the system of names has something to do with this.

However, relying only on names to differentiate people and carry out public management is still crude, and governments cannot quickly differentiate and confirm a single identity among a large group. Even if traditional societies had carried out the simple step of establishing individual names, recording them still had to be done manually. That recording process would be costly and inefficient. For that reason,

---

[6] Regarding the lack of surnames for Turks, Filipinos, and French Jews, as well as other details in this section, see: *Seeing Like a State: How Certain Schemes to Improve the Human Condition Have Failed* by James C. Scott (Yale University Press, 1994).

those societies were still low definition. A villager with a name and a surname could be seen clearly, but once you zoomed out to the city, province, or state, definition would drop. As you try to view a society from higher and higher vantage points, the information about the individual becomes increasingly vague and easy to falsify. The wider the scope, the lower the definition of the information.

A low definition society leads to opacity in the carrying out of social management. This kind of society is cloaked in multiple layers, like a stage behind thick curtains. Behind those curtains, all sorts of conspiracies can be hidden. This allows for the sort of corruption and abuse of power that we have seen repeated ceaselessly throughout history like a chronic and incurable illness.

Disaster relief became a major problem contributing to the instability of past dynasties. When the people were in need, officials had to act, but some of the most serious cases of corruption happened in the course of carrying out disaster relief. The reason for this was that local officials could take advantage of the ambiguity of records to falsify or forge registers of citizens affected by disasters, then request large sums of money, food, and building materials. The central government had no way to verify their claims. Many imperial dynasties attempted to close this loophole by strengthening their official supervision apparatus. By the time of the Qing (1644–1911), the government had set up an authority for inspection of disaster areas and affected households, as well as distributing relief. Despite this, the greatest case of corruption occurred in Gansu under the reign of the Qianlong Emperor (1735–1796). Local officials colluded to receive 2.8 million taels of silver in relief funds. The fraud involved 113 officials and ran for seven years. According to the law of the time, anyone convicted in a case of corruption involving more than 1,000 tales of silver would be put to death. The list of officials to be executed reached 66. The bureaucracy of the region was hollowed out.[7] The Qianlong Emperor was fairly lenient in this case, since he couldn't bear to execute so many officials in a time of relative peace and prosperity.

Unable to fully grasp the truth, humankind faced with their own confusing societies and with nature could only sigh and say: The Dao that can be described is not the true Dao. It is hard to predict what will happen, they said. There are many Chinese idioms that say this. One goes: Grasping human nature is like trying to glimpse a flower through a morning mist. They had to be content with a superficial understanding. Getting close had to suffice. Traditional Chinese philosophy is full of this sentiment and it infused the national culture. This approach to ambiguity gave birth to the mainstreams of East Asian metaphysics.

Behind this approach to the world was powerlessness. As we discussed earlier, industrial society was an identity society. Anyone entering industrial society needed to be able to establish their unique identity. The need to differentiate between individuals became pressing. We already saw the example of governments maintaining photos of individuals in their records. Another example: once people had fixed names and surnames, many states began issuing identity cards that included a unique number. That brought society into higher definition. The Amish never progressed

---

[7] Tang (2015).

beyond that point. They didn't keep up with progress. Refusing electricity is a superficial sign of their outlook on modern life; refusing social management and organization is the real proof of their backwardness.

From the agricultural era to industrial civilization to our present digital age, our image of society is increasing in definition. If Qianlong had the resources available to a modern government, the scale of fraud in Gansu could never have reached such an astounding scale: the government would be capable of quickly confirming the number of victims and they would have been able to confirm conditions on the ground with satellite photography. There's no way to avoid fraud, but it can be ameliorated.

The birth of the super file: single particle governance

The previous chapters discussed the idea of individual pricing—for each customer, you offer a unique price. This is a particularly important characteristic of new economy business models. They are based on insights into the consumer. That insight is gained by constant data harvesting.

Sesame Credit uses a similar individual pricing system. It provides loans to customers on the basis of credit scores, which formerly relied only on data from Alibaba platforms. In order to improve the accuracy of those scores Alibaba began obtaining data from other sources. That was done through business agreements, data swaps, or sometimes cash purchases. The new data was linked to user identities. This was not a simple process, requiring organic linkages based on common characteristics in user identities.

Driven by corporate interests, this type of data integration has ramped up at a shocking rate. According to publicly available reports, only 20% of the data being used for Sesame Credit data scores is drawn from Alibaba platforms.[8] In other words, Alibaba has purchased external customers data that exceeds its own internal data by four times.

The purpose of this data integration is to create a super file for each user (Fig. 4.3).

The so-called super file is intended to constantly collect and integrate a user's data, intending to form the most complete picture of them. A company can cooperate with many other firms, sharing a user's clicking, browsing, and search data from their own site, and receiving that information from sites not under their own corporate banner. These websites might be offering the same products or services, or they might offer upstream or downstream products and services. Apart from that, they can also take data from Weibo, WeChat and other social media platforms. That data can be matched to their existing data using a phone number to confirm user identity.

Once a company starts matching disparate data streams with a single user, the possibilities are endless. Most people use their phone number or email to register for Weibo, so a credit card company can use that information to match social media profiles to their own customer profiles. Once they do this, they can begin to monitor their customer's social media behavior and tailor push notifications. If a customer announces an impending wedding or the birth of a child, new marketing opportunities are opened up. Marriage means buying a house, buying a car, and planning a wedding,

---

[8] Wang (2016).

Fig. 4.3 Three types of data that industry are setting out to connect

and babies need diapers, a crib, powdered milk, and clothes... After a year or so, the kid will need a stroller. When the child is three, they might need books. When they're four, it's time to start planning for kindergarten, and then elementary school. Basically, the child will have different needs at each stage of its life. It goes beyond what I've just listed. There are specialized researchers working on how to market based on the data that they integrate into their own customer profiles.

A piece of information or data can be converted into a lifelong marketing opportunity. There's no doubt that companies will flock to this.

Basically, nearly all B2C companies are attempting to create user super files that can integrate data from numerous sources. The data points that link the data are identification numbers, phone numbers, email addresses, and other unique identifiers. This looks a bit like a tanghulu, the popular Northern Chinese snack, which has a row of haws fruit stuck on a wooden stick. Each piece of data is like one of the haws on the stick.

Although the integration of data is invisible and seamless, and the user will never realize what's happening, it's happening every day. The desire of company's to integrate all data is very strong. In the agricultural era, farmers fought over land, and a similar battle is now taking place in the digital industry over data. Farmers could go in front of a judge to explain their case. "I'm not being greedy," the farmer might plead, "but I want to connect the pieces of land I own." The question is: how far do you go to allow the connection of those separate tracts of land? This request could be extended limitlessly. The same could be asked of today's digital companies. They might say: "I am not being greedy, but I want to connect the data I own." Just like with land, requests to link data could be extended limitlessly.

Companies want to integrate user data with a super file, but governments could do the same. Different departments within a government possess different data on individuals, so it makes sense to link them together in a super file. A citizen's educational, medical, transportation, welfare, tax, criminal, and commercial could be threaded together. This super file will sum up a person's life, from cradle to grave. The database will continually collect and organize new data. Governments can rely on this data

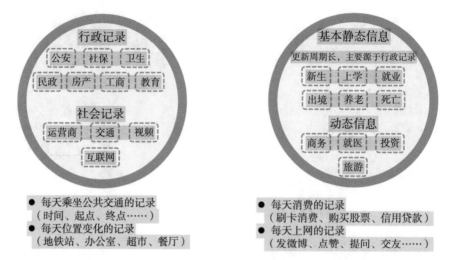

Fig. 4.4 The structure and data required to build a super file

and AI to carry out lifelong recordkeeping, lifelong analysis, lifelong management, and lifelong service (Fig. 4.4).

In other words, a super file can allow governments to carry out analysis and management with the same detail as companies carrying out individualized pricing. In the context of government analysis and management, I call this single particle governance.

The principle of single particle governance is based on using unique patterns in an individual's data. The Chinese word for "principle" is li; according to the *Shuowen Jiezi*, a Han dynasty dictionary, the word originally meant "polishing jade according to an analysis of its patterns." A piece of jade or stone should be polished or cut according to its own patterns and lines. By extension, this means that the patterns and lines of a piece of jade or stone must be analyzed before making changes in a harmonious and orderly fashion.

With the foundational goal of bringing society into high definition, single particle governance using super files can make detailed, individualized management a reality. One example of this is in taxation: in a cruder national system, taxation can only be divided based on income brackets. Each tax bracket pays a different proportional rate, but it doesn't take into account things like how many dependents they have, number of children, or cost of living. Those things are outside the purview of a basic tax management system. It's unfair for the head of a large family and a bachelor to pay the same tax rate. In countries with a higher degree of management, more data points are used to decide taxation more fairly.

Since 1975, the United States has had an earned income tax credit (EITC). Every household pays taxes based on a number of factors, such as the income of a couple, number of children, residency, etc. Taxpayers get tax credits as a refund, based on their particular circumstances: if they have a higher burden, they receive more in tax credit refunds; if they have a lighter burden, they receive less.

Digression: China's own examples of granular, individualized social management

Since 2017, many cities in China have introduced aggressive policies for maintaining human resources. The goal is to attract talent, create talent, and maintain competitiveness. Carrying out a scheme to attract and foster talent requires more than general data; a government must carry out a detailed cross-analysis. They must know, for example, how many graduates of local universities get jobs in the area, how many are from outside the province, how many are male, how many of them have bought a house within three years, etc. The information from this cross-analysis can be further partitioned. Once you do that, you can track categories like people from out-of-province, engineering students, men, graduates of out-of-province universities, or home buyers, figure out why they have decided to stay, then design schemes to keep or attract more people of those categories. By recognizing local human resources needs and the concerns or anxieties of the talents you hope to attract, a more goal-oriented solution can be proposed and management can be more granular.

Another example is the "targeted poverty alleviation" carried out in China since 2013. The Chinese government set a goal of bringing 70 million people out of poverty by 2020. "The value, the effectiveness, and the success of these programs depended on their precision. This meant accurately understanding the situation of poor households and taking targeted, individualized, and granular measures to provide assistance.

Generally speaking, making these classifications more detailed and individualized is not simple; it's more than a matter of making an either-or decision about whether a person or household belongs in a particular category. Rather than drawing crude lines around names on a list, detailed classification would require more complicated diagramming, with a gradual, intersectional, overlapping appearance. In the language of data analysis, the refinement we're talking about is infinite partitioning and crossover analysis. Lacking enough information and data points to differentiate and describe individual circumstances and ambiguous conditions, the analysis cannot be fine-tuned: it's like a hand without thumbs attempting to turn a screw.

At this point, I've reached a conclusion: if a country wants to manage society, it must invent an individual unit that can be clearly inspected and identified. Bringing society into higher definition is a prerequisite for effective governance. Bringing society into higher definition is the basis of refined, individualized, and intelligent management.

Single particle governance will take as its subjects not only the consumers that we have talked about at length, or citizens, or vehicles, but also other objects.

Express courier services have already realized a system of single particle management: in the process of shipping an item, each step is recorded in real time, from pulling it off the warehouse shelf, to dispatching it, to transferring it between delivery

Fig. 4.5 A high definition
society is the foundation of
refined, individualized,
intelligent management

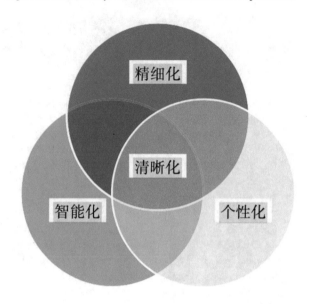

services, to taking it to a customer's door, to the final receipt... The majority of express courier services give customers a platform to check the status of a delivery, whenever they like.

We should be careful about using the word "item," though, since it's slightly ambiguous. Take a bottle of water as an example... Pull two bottles from the same box and they become a quantity of the same item, rather than two separate items, since their contents and packaging are identical. There is no system to differentiate the two bottles.

But there will be (Fig. 4.5).

In January of 2014, Alibaba spent around $170 million to buy a 54.3% stake in CITIC 21CN. Apart from CITIC 21CN holding the sole license to sell pharmaceuticals online, Alibaba also took note of the company's unique electronic drug monitoring code system.[9]

In the past, pharmaceuticals could only be grouped by item. Just like the bottle of water we used in our last example, there was no way to differentiate two identically packaged bottles of pills. The goal of an electronic monitoring system was to give each and every pharmaceutical item a unique code. With that code, there could be real time tracking and management from the point of manufacture to the point of prescription at a drugstore, hospital, or pharmacy.

In other words, using the unique code on a box of medicine, a customer could use their phone to get information on where it was produced, how it was shipped,

---

[9] On February 20th of 2016, the State Food and Drug Administration suspended the implementation of the electronic drug monitoring code system, withdrew Alibaba's operating rights, and announced that they would be seeking a third party operator to implement the system. In April 2016, Hong Kong's Securities Regulatory Commission determined that Alibaba's investment in CITIC 21CN violated their regulations.

and where it was sold. It would also record who purchased the item. If there was any problem with the medication, the regulatory department could quickly determine where it was manufactured, where it was sold, and who it was sold to. In the event of a recall, there could be absolute accuracy.

As this technology develops, this kind of single particle granularity might be applied to managing the manufacture, distribution, and sales of other commodities, such as food. Existing technologies, like radio-frequency identification (RFID) could be used for this, as well as QR codes. Ten times the information of a traditional barcode can be encoded in a QR code, and they can be embedded in video, images, and sounds. This is a low cost way to embed a lot of information in a way that consumers can access easily. The product and its data will be linked, allowing for trajectory management.

Our modern societies are complex systems, involving countless elements linked together. But the key to social management comes down to people, then the vehicles and other objects they interact with. These things are the lifeblood flowing through society. Single particle governance is an attempt to train a magnifying glass or a microscope on society, people, vehicles, and objects. These are infinitesimally tiny particles that make up society; single particle governance attempts to separate them out for inspection. It could be compared to looking at an ocean: we can only see the waves on the surface and the direction of the tide, but single particle governance is like being able to analyze a single droplet of seawater.

This ability is like being able to lock in on a snowflake in a blizzard. Snowflakes all look the same, but the process that forms them gives them each their own unique structure. They each fall in a unique path, too, following air current. Each snowflake is different and moves in a unique trajectory. But our technology at present is like seeing through the eyes of God: we can lock onto, track, and differentiate each snowflake as it falls. This kind of power has been possessed by no government before in history.

As we have already seen, Internet companies now have many platforms for e-commerce, social networking, transportation, dining, logistics, etc. With the power of granular management, we are at the point where governments will develop a similar platform. The mechanisms involved will be similar to those of Internet companies, but they will have a deeper and wider source of data to draw from.

With this platform, governments around the world will have amazing new capabilities in governance. This is already becoming a reality and we are already seeing small displays of new capabilities, which should be able to solve problems that hitherto seemed insurmountable.

This new platform is also gradually producing some new problems. This will present a problem for pure pragmatism. If these new powers are used inappropriately, they will create a data cage for citizens. This will turn the platform into the Panopticon that Jeremy Bentham (1748-1832) proposed, where one watchman can observe everyone.

As we enter the age of single particle governance, the first to be tossed behind bars will be human nature. That holds true for both democratic and autocratic regimes. Our understanding of human nature is only general. It's invisible and imperceptible. But that understanding contains contradictions. A new tension has appeared. As data rises, human nature recedes.

4.2 Don't Leave it to Luck: The Experience of Governance in Ancient China

> Heaven can be guessed at, Earth can be measured. It is only a person's heart that cannot be protected. But to see loyalty crimson as blood, who knows what false words might be spoken as fluently as someone plays a flute. They urge you, my lord, to do this and then do that, or your wife and you will be as distant as the morning and evening stars. They urge you, my lord, to catch a bee, then say you must not catch it, or you and your son will fight like jackals and wolves. A fish at the bottom of the sea and a bird in the sky. As high as it is, it can be shot down. As deep as it is, it can be hauled out. When people's hearts and minds are in opposition, no matter how close they are, they cannot be understood.

This poem by Bai Juyi (772–846) talks about the complex, changeable, elusive character of human nature. We have compassion, shame, respect, conscience, and we also have joy, rage, likes and dislikes, and various appetites. Human nature is not only changeable, switching between love and hate, or good and evil, but it can transform as fast as thought. This is the thinking behind the idiom that goes, "A wicked thought follows a good thought; a devil in one moment and a saint in the next."

What can we use to restrain this instantly changeable human nature? Ethics and law attempt to answer this question, but in a low definition society, they lack precision. We can see that not all violations of ethics and morality end up being punished; the punishment is not always timely; and the punishment might not always fit the crime.

It's from this that we get the human tendency to trust in luck.

This trust in luck is a challenge to social management. In a country where people trust in luck, there is a lack of self-restraint. The government can only rely on law enforcement and the courts to keep order and stability. This leads to a large number of people becoming two-faced: they behave themselves when they know they're being watched, but not when they think nobody has eyes on them. The crime rate goes up and there can be a breakdown in the normal functioning of society.

This challenge runs through modern human society. Societies in the East and the West have thought about it differently, and they have provided different solutions, reflecting differences between the two civilizations. Eastern cultures relied on a belief in the original goodness of human nature to establish the four cardinal virtues of humanity, justice, propriety, and wisdom; Western cultures' belief that men are born wicked shaped their creation of laws and charters.

Digression: How giving up on luck can be an important step in self-improvement

The Buddhist monk Hong Yi (1880–1942) believed that good luck was a long term detriment to personal development. For example, if you get away with making an unkind remark, you might do it again; if you break the law to earn a profit and get away with it, you might do it again. This sounds like good luck, but it actually leads to disaster and suffering because they don't teach us how things actually work. Luck is not in accordance with the objective rules that govern the world. If you draw conclusions from your good luck and hope for more of the same, you will certainly be disappointed.[10]

Zeng Guofan (1807–1872) held a similar point of view. After some setbacks in his career as a bureaucrat, other people blamed it on his bad look, but he disagreed. He said that some officials received praise for their writing when it was undeserved, but no truly fine writing had ever been completely overlooked.[11] He vowed to improve the level of his own writing through hard work. Zeng Guofan was an orthodox Confucian, so he was familiar with the Master's lines: "To give oneself earnestly to the duties due to men, and, while respecting spiritual beings keep aloof from them, may be called wisdom." But he often included in letters to his younger brother an admonition to keep an eye on spiritual beings, since they know everything we think, do, and say; we can fool people, but we might not be able to fool the gods. What he was telling his brother was that he had to work honestly and work hard, never relying on a good opportunity to come along or good luck to save the day.

Past generations of statesmen have attempted to dislodge the belief among their subjects in luck. We can see that in the selection from the *Commentary of Zuo* that begins the chapter: if a country's common people believe in luck, it will be bad for the nation; if a country is ruled wisely, then the people will know that their good fortune has nothing to do with getting lucky. Xunzi argues in *On Enriching a State* that the ideal society is one in which "no official is promoted without merit; no subjects earn anything through mere good fortune." What Xunzi means is that no bureaucrats should reach their rank without merit and hard work, and that none of the common people should sit around waiting for good fortune to find them. During a particularly tough time in the Song Dynasty, imperial official Fan Zhongyan (989–1052) took part in the Qingli Reforms, presenting his own ten-point proposal for reforming government. The second of those proposals was to suppress belief in leaving things up to good fortune.

The reason that it was so high on the list in a discussion of national policy is that politicians had realized this way of thinking could be a disaster. Of course, there were examples in history of people doing bad things and getting away with it, but it isn't good for this mindset to become consensus.

---

[10] The original quote is: "The most unfortunate thing in life is when one makes an indiscreet remark and gets away with it; when one fails to carry out a plan but suffers no setback; when one puts in no effort but gains reward. Later, one will begin to perceive this turn of events as normal and is not disturbed by it. This is a tragedy, and disaster will come from it." *A Record of Master Hongyi's Maxims.* (1997, May). Anhui Literature and Art Publishing House.

[11] The original quote is: "Inelegant writing may earn temporary reward; no excellent writing will be buried forever." *Zeng Guofan's Collected Works: Letters Home (First Volume)* (2001, September). Jinghua Publishing House.

Let's look at some examples from soccer crowd management. The cost of organizing and keeping order at a match is extremely high. One of the reasons for this is that so much uncivilized behavior goes on in the stands. People throw things, like bottles, lighters, and coins onto the field, for example. The language used by audiences can be inappropriate or provocative, leading to mass disturbances and riots. No matter where you are in the world, soccer matches have a heavy police presence. According to statistics on soccer security in the 1990s, Sweden was spending 7.5 million Euros a year, while the top division in Italy was dropping 40 million Euros. The reason for the behavior of soccer fans is that they are in a crowd. No matter how many police are there, it is difficult for offenders to be held accountable.

Since the advent of surveillance cameras, the situation has improved markedly. Sweden was among the first countries to introduce the new technology, installing cameras in three major stadiums in the 1990s. Unruly and illegal behavior was reduced by 65%. After that success, the Swedish Football Association made a rule requiring any stadium that hosted league matches to install surveillance cameras.[12] This initiative cut down on the cost of policing.

The ubiquity of surveillance has uncovered some unbelievable things happening at stadiums.

At a baseball game in Atlanta in July of 2015, two sisters were eavesdropping on the text messages of a woman seated in front of them. The woman had her husband beside her, but unbeknownst to him, she was texting the man she was having an affair with. The two sisters immediately took out their own phones to document the text messages. When the game ended, they handed a note to the husband: "Your wife is cheating on you. Look at the messages under 'Nancy.' It's really a man named Mark Allen." She apologized for being the bearer of bad news and offered to send pictures confirming the text messages. Half an hour later, the husband contacted the sisters to ask for the photos.[13]

This is not the first time that an incident has happened. In December of 2014, while a man and his pregnant wife were watching a Detroit Lions game, an eavesdropper noticed that she was sending texts to another man. The eavesdropper handed a note to the husband and later shared the story on Facebook. With cameras everywhere, it can feel like we're constantly being watched from above. Without surveillance tools and methods, it's hard to imagine good fortune ever leading to these affairs being exposed (Fig. 4.6).

We're living in a different age.

We are currently entering an era of universal recording. That means recording equipment will be increasingly small and increasingly common; recording methods will be increasingly convenient; recording will take place everywhere and by many means. In the past, we had to decide what to record; now and in the future, we will have to decide what not to record. With the universalization of recording, everyone is recording everyone else. We're entering a society where nothing is left up to luck.

---

[12] Priks (2014).

[13] Mazza (2015).

Fig. 4.6 A cheater in the stands is exposed

4.3 The Power of the Individual: Everyone Holds a Sword and Everyone Is Held a Swordpoint

Ancient Greek literature is the source of the story of "The Sword of Damocles." In the 4th century BC, Damocles was a courtier in the court of Dionysius II of Syracuse. He was envious of the king. When Damocles flattered him about his immense power, Dionysius II offered to switch places with him for a day. He told him that he could experience what it was like to be king.

At a banquet that day, Damocles was enjoying his time on the throne. He happened to look up, though, and noticed that a sharp sword was hanging over his head from a single hair of a horse's tail. Watching the sword hanging perilously over his head, in danger of falling at any moment, he immediately understood what the king was trying to tell him: with immense power comes responsibility and risk. If a ruler mismanages his people, he could face serious consequences, up to death.

Behind powerful forces there are generally massive risks lurking. These dangers might appear at any time, but there's no way to predict them. This is the power of "The Sword of Damocles." In today's world, anyone can pull out a phone and record what's going on around them, turning a fleeting moment into a stream of data. That data can travel around the world in an instant. This is an era where everyone has a camera and anyone can be a reporter; everyone can be recorded, too, especially public figures. Now that recording is much easier and becoming universal, the individual has greater power; the borders of the individual are expanding; and it means that the individual has more power to counteract the system.

This is like everyone holding the sword, but also sitting under the sword. This will not only change the behavior patterns of individuals, but also social management.

In March of 2012, Evergrande chairman Xu Jiayin attended the National People's Congress and People's Political Consultative Congress (the Two Sessions) of the Chinese People's Political Consultative Conference (CPPCC) as a delegate. He did so while wearing a Hermès belt. Photos of Xu Jiayin began circulating online and social media users began digging into the extravagant wardrobes and accessories of Two Sessions attendees. They discovered one representative in a brand new Pucci suit, another carrying a 100,000 yuan Hermès handbag… This led to lively discussion online. After this incident, the delegates quickly adjusted their behavior. As far as wardrobe and accessories, the Two Sessions returned to being a simple, low-key affair. Luxury goods, like fur coats, designer bags, belts, and glasses stopped being worn by Two Sessions attendees.

"Brother Wristwatch" Yang Dacai might be the best proof that the space for leaving it to luck has shrunk. On the 26th of August, 2012, a major traffic accident on the Baotou-Maoming Highway resulted in 36 deaths and Yang Dacai went to inspect the scene in his role as head of the Shaanxi province work safety administration. His cheerful demeanor in photographs at the accident scene was roundly condemned online. Social media users went a step further, starting up one of the online vigilante campaigns colorfully referred to in Chinese as a "human flesh search engine." Yang had first been nicknamed "Director Smiles," but his nickname was changed to "Brother Wristwatch" when his collection of expensive watches was uncovered by browsing previous photo ops. He had worn at least five different, highly valuable watches. Even more suspicion fell on him.

Yang Dacai was forced to respond. "In the past ten years," he said, "I have indeed purchased five watches. I bought them myself, out of my own salary. All of this has previously been reported to the Party disciplinary supervision department."

His response prompted even more interest. Social media users turned up photos of Yang Dacai with a sixth watch, then a seventh. That continued until 11 watches had been found. Professionals set to work identifying them and putting prices on them. The watches included pieces from Rolex, Omega, Bulgari, and Vacheron-Constantin, valued in the hundreds of thousands.

The story of Yang Dacai shows us that even a fragmentary piece of data has a lot of potential energy. He was not only raked over the coals and given the nickname "Brother Wristwatch" but also faced investigation by provincial Party discipline officials. Yang Dacai was expelled from the Party and his case was transferred to the courts. In September 2013, he was found guilty of accepting bribes and possessing large amounts of property that he could prove the source of. He went to prison for 14 years.

Having learned from the situation of "Wristwatch Brother" Yang Dacai, some other officials stopped wearing expensive watches or flashy accessories in public. At a media training event in Foshan in 2013, one of the experts advised, "It's now best to avoid a necklace or chain, too, since people are looking for those, as well as watches." Sharp-eyed social media users began catching tan lines on the wrists of officials, a sure sign that they had ditched their watches just ahead of taking the podium at a press conference.

The fall of Yang Dacai is a landmark event in public participation in social governance. Right of supervision is granted to citizens by law, but, since most people have no idea how it might be exercised, it mostly remains academic. The spread of recording technology means that citizens can learn the facts about people and events. This is an ability that has been entrusted to everyone.

In July of 2016, the Ministry of Public Security affirmed this right of supervision, telling its officers to get accustomed to public observation and being filmed while doing their jobs. In September of 2016, the Ministry of Public Security issued their "Provisions Concerning the Collection, Extraction, Review, and Judgment of Electronic Data in the Handling of Criminal Cases," providing law enforcement with the authority to use electronic data in court. This data could include web pages, blogs, Weibo posts, WeChat Moments, photographs, music, etc.

Chinese police will eventually get used to policing under a lens. This is an example of progress in social governance thinking and will provide new solutions to long-standing issues. As networked devices and the Internet penetrate fully, and mobile phones and surveillance cameras become ubiquitous, countless details about the operation of society will be recorded; the role of data will become increasingly important; and these situations will become the "new normal."

4.4 Data Is Incontrovertible: A World Where Nothing Is Left to Luck

As we have already seen, data is pressing down on human nature, challenging its tendency to leave things to luck. If people give up on the psychology of flukes, the world will be a safer and more peaceful place.

In August of 2016, Gao Chengyong was arrested in Baiyin, Gansu and charged with the vicious murder and rape of eleven women in Baiyin and Baotou.

Prior to this test, although the police had repeatedly locked down thh area around crime scenes, Gao Chengyong had evaded capture. I noted that what allowed the police to nab Gao Chengyong was the use of a Y-chromosome DNA test. This is a test that can determine the paternal lineage of a person, as well as revealing other paternal kinship relationships. Y-chromosome DNA allowed the police to link a suspected serial killer to a member of Gao Chengyong's family in their custody for unrelated crimes. That allowed them to quickly narrow down the list of potential suspects. When Gao Chengyong was captured, the police were able to carry out other analysis to determine he was the culprit. Cases that had been unsolved for years were suddenly broken wide open.

Looking over the history of the fight against crime, we have no choice but to admit that limited investigation techniques often led justice to be served late or not at all. However, as big data technology advances and every step we take is increasingly recorded, there will be a revolution in the domain of public security. As I see it, police work will be increasingly dominated by data collection and analysis.

The Baiyin serial killer case is not an isolated case. A similarly legendary example of police work took place in 2016 in Hangzhou, where I have been living for the past several years.

The Zhijiang Garden Murders took place in 2003. A man surnamed Yu entered a house, killed three occupants, and then disappeared without a trace. That year, the

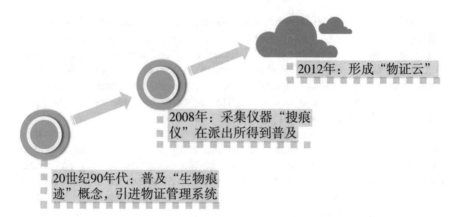

Fig. 4.7 The "Evidence Cloud" of Zhejiang is already sharing data across multiple jurisdictions

legendary Chinese-American forensic scientist Henry Lee visited Hangzhou for the first time and was asked about the case. He had no solution, but said this: "Give it enough time and the case will be solved."

Some might sense a certain helplessness in this remark, but Henry Lee was aware that technology was making progress.

In the 1990s, Hangzhou police started collecting "organic evidence" and introduced a management system for it. In 2008, this collection and management system was spread to police stations across the city. It was capable of collecting images, DNA, fingerprints, palm prints, footprints, and shoe sole patterns. In 2012, all of the data in this system began to be transferred to an "Evidence Cloud," so that any suspect's data could be integrated for large-scale comparison (Fig. 4.7).

In September 2015, at a restaurant in Zhuji, about 40 miles south of Hangzhou, a man was arrested over an argument that led to a stabbing. A DNA sample was taken by local police. With that DNA information on an interjurisdictional "Evidence Cloud," Hangzhou police were able to determine that the man from Zhuji was the same man that had carried out the Zhijiang Garden Murders. The identity of the mysterious Mr. Yu was confirmed.[14]

You might say they got lucky. But solving the case became inevitable after the technology was popularized.

Again, the solving of the Baiyin serial killer case and the Zhijiang Garden Murders are not isolated incidents. They are part of a trend in policing that has also been seen in the United States. In the first half of 2018, American police cracked a cold case that was quite similar to the Baiyin serial killer.

Between 1975 and 1986, a sick killer had stalked California. He was suspected in at least 12 murders and 45 rapes. Investigators chased the "Golden State Killer" for twenty years, looked into thousands of suspects, but came up with nothing.[15]

---

[14] Chen (2016).

[15] Haag (2018).

In December 2017, a detective in California came up with a new solution. The killer's DNA was on file, so he decided to upload it to the genealogy website GEDmatch. This is a platform that analyzes genetic information and has helped people find shadowy ancestors or lost relatives. The website quickly turned up some genetic matches for the killer. Having made this important discovery, the list of suspects was narrowed from thousands of names down to one family. Investigators further narrowed that down to one person: 72-year-old Joseph James DeAngelo. In April of 2018, the "Golden State Killer" was finally brought to justice.

This is almost the same as the situation in Baiyin, but there's another American case that looks even closer to what happened in Hangzhou.

On July 7th, 1991, police in San Bernardino, California found a body stuffed under the mattress. They determined the identity of the victim and also picked up a fingerprint from their car. No match was found in the local police database. The case went cold for two decades.

Beginning in the 1970s, American law enforcement began building an Integrated Automated Fingerprint Identification System (IAFIS). This is a system to match fingerprints taken across the entire country. It built up a database over three decades and has the fingerprints of 70 million criminals and 40 million ordinary citizens, as well as images of faces, scars, and tattoos. IAFIS data is constantly growing and improving.

In 2010, the fingerprints from the 1991 San Bernardino murder case were entered into IAFIS. After five hours of analysis, the system spit out a list of suspects. One of them was living in Tennessee, thousands of miles away from San Bernardino. After further manual analysis and interrogation, Arrowood confessed. In 2014, he was convicted of first degree murder and sent to prison for 25 years.[16]

The Arrowood case and the Zhijiang Garden Murders case both show the use of interjurisdictional data connectivity. You can imagine how many cases went unsolved in a time before law enforcement data was connected. If a murderer got away with a crime by good luck, justice would never be served.

That's another thing that links these two cases. After the murders, both criminals ran far away. That's typical for criminals everywhere.

Where do they run to? In former times, criminals could run to the wilderness and take refuge in a temple. Their remote location made Buddhist temples and Daoist shrines ideal for this. They became known as a places that "sheltered evil people and countenanced evil practices." There is another famous phrase: "On the endless sea of suffering, repentance is the only harbor; if you reform an evildoer, you reach Buddhahood instantly." This suggests something about the easy relationship between religious sites and criminals. Of course, heroes might also hide out in those places, but they kept a low profile, too. It is because of a lack of data or records that no one could easily know the situation or history of people in hiding.

However, nowadays even some monks find themselves caught up in the data net. In 2012, police at Haozikou in Neijiang, Sichuan began carrying out the "Standardized

Address and Three Facts" program, recording addresses and other census information from residents. Police were surprised to learn that one of the monks in the jurisdiction was the suspect in a fraud case. In 2016, police in Fengyang, Anhui discovered a wanted criminal through facial recognition and were amazed to discover that it was Zhang Liwei, head of a local temple, vice chairman of the local Buddhist Association, and member of the CPPCC. He had committed a triple murder in the Northeast in 2000 and run off to Anhui. After becoming a monk, his life had gone smoothly, and he was riding the crest of success.[17]

In a time without data, there's no way to check a person's background. When we had less data, a person's background could be selectively recorded on paper, but it would often be left moldering in an archive. Nowadays, there is a record of most things we do. That record is far more durable than paper records, and it's stored in the cloud. At any time, it can be analyzed again and integrated with other data. Data is like a lie detector. Any data that comes up as "irregular" can prove an illegal act.

I worked for eight years in public security for a ministry in charge of border defense. I investigated maritime criminal cases. Ten years ago, if someone with a small craft committed a crime, it was very easy for them to escape punishment by simply moving to another jurisdiction. Last year, a former coworker let me know that this method of escape is no longer feasible. Nowadays, all provincial data is online. If the suspect moved from Shanwei to Shantou, the police in Shantou could quickly determine their identity and force them to pay their fines.

When data about an individual is isolated, that individual's behaviors are also isolated. If someone is good at disguising their identity, they can quickly become a completely new person. When data is connected, the individual's behaviors are also connected; data and behaviors are mutually connected and provide mutual confirmation. Police investigation and interviews are all about collecting data, connecting information, and finding inconsistencies.

Digression: how to use data to uncover money laundering

During my work at Alibaba, our connections with the government usually came through law enforcement departments. The Public Security Bureau often hoped to obtain transaction data from Alibaba platforms as a way to confirm evidence or seek new evidence. Transactions on e-commerce platforms are data-based, so they can be quickly tracked. Beyond simple tracking, the transactions can also be analyzed and their patterns used to make forecasts. Anti-money laundering experts can identify signs of an illegal banking exchange: an account receiving and sending money to a large number of other accounts, the addresses linked to those accounts have a large geographic distance between them, many transactions in a single day, consistent deposits and intermittent payments, and the sums of money being transferred divisible by the daily exchange rate. At that

---

[17] Chen (2016).

point, further manual monitoring can be done. In the first half of 2017 alone, the data analysis center of Ant Financial referred more than 300 suspicious transactions to the Public Security Bureau. The space allowed for economic crimes, such as money laundering and tax evasion is continually shrinking.

Data is incontrovertible. By making wise use of big data, we can enter a more secure era.

According to statistics, the detection rate for homicides by Chinese law enforcement exceeded 95% over the past five years. Compared to numbers in 2012, the number of serious violent crime cases nationwide in 2016 dropped by 43%. In August of 2017, the government of Zhejiang held a press conference to announce that their homicide detection rate was above 99%. In the first half of 2017, the number of public security cases in the province dropped 7.76% year-on-year; criminal cases dropped by 28.59%. My friend at the Public Security Bureau explained it like this: "Big data and new tech is so powerful. If we get a case, we solve it. After that, we can start working on cold cases. We're either solving cases or waiting for cases."

In the past, when darkness covered half the Earth each day, no electric light disturbed it. When the sun set, perverts and criminals came out to stalk their prey like wild beasts. Traveling at night was a risk. There was a good chance that you would lose your possessions, if not your life. The philosopher Thomas Hobbes (1588–1679) once wrote that what he feared most was lying alone in bed at night. He wasn't afraid of ghosts. He was worried someone would come in, smash his head in, and take his money. A dark night is the perfect cover for crime. "At night," an Italian proverb tells us, "the housecat becomes the leopard." Historical data proves this to be true: in a time without electricity, people came out at night, and 75% of thefts took place after sunset.

At the beginning of the 19th century, Europeans invented the gas street lamp. This was a turning point in human security. In 1823, 40,000 street lamps illuminated more than 300 kilometers of London streets. Other cities soon followed their example, including Paris, Berlin, and Boston. Whenever there was a riot in a city, the first target of bad elements was always the street lights. Jane Austen (1775–1817) once observed that gas street lamps had done more for crime prevention than anything else over the previous thousand years.[18]

In the 1880s, Edison gave us the light bulb and built the first power company in New York. That was the beginning of electrification in cities. Electricity brought a stable and durable lighting source to the city. The sun lit the day; electricity lit the night. The only difference between day and night was the source of illumination. Soon, sociologists began to observe that the crime rate would drop in any part of the city where lighting was installed. This was further proof that criminals were making use of the cover of darkness. Today, data has the same effect. The ubiquity of surveillance cameras and the rapid integration of data are like a new street light,

---

[18] See Jane Austen's *Sanditon*, an unfinished novel begun 1917.

illuminating the dark side of human nature, letting everyone know that they can't leave anything up to luck. Data is giving birth to a safer society. Our generation will witness human history's most substantial and complete decline in crime rates.

References

Chen Leideng. "4,600 days after the Zhijiang Garden Murders that shocked Hangzhou, the case has been solved." (2016, June 11). *Qianjiang Evening News.*

Chen Nuo. "During his 16 years on the run, fugitive in murder case laundered his identity by becoming the head of a Buddhist temple." (2016, August 19). *Xinhua.*

Haag, M. "What we know about Joseph DeAngelo, the Golden State Killer suspect." (2018, April 26). *New York Times.*

Lin Ping. (2017, February 14). "Supreme Court: 6.73 million cases of 'trust-breaking,' nearly a million scofflaws forced to fulfill debt obligations." *The Paper.*

Mazza, E. "Cheating wife reportedly busted while at a baseball game." (2015, July 27). *The Huffington Post.*

Priks, M. "Do surveillance cameras affect unruly behavior? A close look at grandstands." (2014). *Scandinavian Journal of Economics*, 1160-1179.

Tang Bo. (2015, June). *Inquiries into the Secrets of the Qing: Anecdotes from the Imperial Court.* Guangxi Normal University Press

Wang Ling. (2016, April 29). "Sesame Credit: Alibaba accounts for less than 20% of data." *Caixin Online.*

Chapter 5
The Force of Data: How Universal Recording "Warps" Human Nature

Abstract Humans and animals each have their own respective strengths. Wind and fire are powerful, too. Everything has its own energy inside of it. Is data an exception? This chapter attempts to define "data force" and lay out how it affects human nature. As we already know, a force can be intangible; we know that an object can exert force on another object without coming into direct contact with it. Like gravity, magnetism, the attraction that holds particles together, and centripetal force, data force is also invisible. If human society is a vast river on which we all float, there are two ships leading our armadas: the first is reputation and the second is benefits. The beginning of this book focused on the latter, explaining how the production and use of data can potentially be of benefit to the individual. This chapter focuses on the former, discussing how data force influences humanity's quest for fame and good name.

5.1 In All of Our Science the Pursuit Least Developed but the Most Valuable is the Study of Mankind Itself

As I wrote earlier in the book, the digitization of everything is allowing us to bring society into higher definition, making possible single particle governance; universal recording will hold people accountable for their behavior, so they will learn to regulate it; and we have seen the possibility of a "world without thieves," where people believe that social management will win out over their own good luck.

Long before the digital age, we already had a tradition of recording things. The first tool was writing. Talking about forces, there is certainly a force present in the written word that can act on human nature.

In Chinese history, recording events was once regarded as something holy. Ancient men once said, "Understand the past, to draw lessons for the future."[1] There is a wealth of knowledge and experience contained in historical records; we must use history as a teacher. Whether we are talking about the thinkers of the late Zhou or the rhetoricians of the Spring and Autumn and Warring States periods, when it came time to argue their case to the elites of the nation, they worked from historical events.

---

[1] This is from a Tang history of the Six Dynasties.

© China Translation & Publishing House 2022
Z. Tu, *The New Civilization Upon Data*,
https://doi.org/10.1007/978-981-19-3081-2_5

The words and actions of ancient sages formed the orthodoxy of every philosophical school. Anybody that wanted to be thought of as a learned man could not afford to be ignorant of the historical records. This gives us the Chinese idea of *fagu*, or "imitating the ancients."

The record of past events became *shi*, or history. A position was created for what was called *shiguan*, a term that goes beyond mere historian, describing an imperial official whose job it was to maintain and compile historical documents and records. Among ancient civilizations, China put the most emphasis on the study of history. "Among all academic disciplines in China," Liang Qichao (1873–1929) said, "the study of history is the most advanced. And among nations studying history, China is the most advanced."

Before the *shiguan*, there were shamans occupying a similar position.[2] This is more proof of the quasi-religious nature of the official historian's role. They also held a monopoly over this power to record: beginning in the Sui (581–618), it was forbidden for anyone outside of the official historians to record dynastic histories or comment on historical figures.[3] The right to record became indivisible from the authority of a monarch.

Recording events is not a passive act. It's a generative act. It turns events into data. That data is dynamic. I call the power contained in data a force. Like the force of gravitational waves, we can prove its existence, even if we can't see it. Data force can influence and change human behavior and tendencies, and it has had a role in and even created many great changes in human history.

As we saw in the last chapter, humanity is moving toward an age when nobody believes good luck will get them out of a jam. It's going to be a safer age because of the deterrent ability inherent in universal recording. This all comes down to data force. This chapter focuses on how this data force will affect the human desire for fame and reputation.

Humanity realized early on that their lives were limited. The only way to span time was through achieving posterity—having your name and deeds recorded in history. Confucius said: "A man of noble character hates the idea of his name not being mentioned after his death." Ouyang Xiu (1007–1072) said this: "Those whose deeds are recorded in books shine as brightly as the sun and stars." The meaning of these two quotes is that right-thinking people should be concerned about posterity; and the only way to be celebrated posthumously is to have your name written in the history books. The Tang Dynasty history Liu Zhiji (661–721) explained this very clearly:

---

[2] The term *wuzhu*, or shaman, in ancient times referred to those involved in the supernatural, including presiding over religious ceremonies and funerals, and later to soothsayers.

[3] The original lines, taken from the *Book of Sui*, the official history of the dynasty, composed in the Tang, say: "Common composition of court history, passing judgment on historical figures is forbidden".

Our lives are balanced between heaven and earth. We are like mayflies, like the glimpse of a white horse as it runs by a gap in the wall—passing in a flash. We struggle to leave behind a proud legacy but face the possibility of meeting death without having made a name for ourselves. From the emperor on high to the lowliest subject, from the ministers of the court to the men of the mountains, everyone is in tireless pursuit of renown. Why is this the case? Everyone wants immortality. What does that mean? Immortality is having your name passed down through history.[4]

It is in our nature to hope that we might achieve posterity that stretches beyond our own limited lifespans. It's instinctual to want to be remembered by later generations. This is especially true for the ruling class. The early official historians recognized this natural inclination and attempted to construct systemic barriers to entry, allowing their monarch to maintain authority even over posthumous remembrance. This is what is meant by the phrase "governing through history."[5]

In China's long history, the idea of using history to control the present appears and disappears. From this concept, we can deduce the risks of recording history. The instrument of recording history was the brush, but it could be wielded like a blade. Recording history was a matter of life and death. In the background of these records, we know that blood was spilled.

5.2   The Bane of an Emperor's Existence

In 960 AD, the Fine Dynasties and Ten Kingdoms were both reaching their end. This was an era of great division in Chinese history. In that year, troops of the Later Zhou marched out to Chen Bridge in Henan and abruptly declared that military governor Zhao Kuangyin (917–975) was the true emperor. Zhao Kuangyin declined the new position at first, but then accepted it, leading troops back to take the capital. With the military behind him, there was no resistance; the emperor was forced to abdicate. This was the beginning of the Northern Song Dynasty.

Most historians now believe that the Chen Bridge Mutiny was a coup d'etat meticulously planned by Zhao Kuangyin. When he ascended the throne, he became a prudent, effective leader. He was praised by later generations for bringing to an end the chaos of the age, leading to a new era of stability and progress. He became known as Emperor Taizu of Song.

Emperor Taizu was a man of modest temperament, who enjoyed his leisure time. He could probably be called China's first soccer star—he was fond of playing cuju, a game strikingly similar to soccer, with his subordinates. His other hobby was hunting birds in his yard.

---

[4] This is from Liu Zhiji's *Shitong*, or *Generality of Historiography*, composed in the sixth century.

[5] Many historians of the past, including Jiao Hong (1540–1620) of the Ming, as well as modern historians, like Liu Yizheng (1879–1956) believed that official historians had a kind of "historical authority" that could restrict the monarch's power. They called this "governing through history.".

One day, he was out in his yard, hunting with a slingshot. He was closing in on his prey when he heard footsteps behind him. He was told by an aide that a certain minister had an urgent matter to report. Zhao Kuangyin waved him off, saying that it could wait. The aide insisted that the matter was very urgent. Zhao Kuangyin finally tossed aside his slingshot and went back to the court. As soon as he heard the minister's report, he lost his temper, asking why he was being disturbed over such a pedestrian concern. The minister said firmly back to him: "However urgent, I believe any state concern should take precedence over hunting birds."

Zhao Kuangyin lost his temper and roared back at the minister: "If I can't hunt birds, maybe I'll hunt you." He reached for an ax handle that happened to be lying on the table, raised it, and began bringing it down on the minister's face. A pair of busted teeth fell in the pool of blood at the minister's feet. The minister did not cry out, however, and simply bent down to pick up his teeth. That made Zhao Kuangyin even angrier. "What are you picking those up for?" he demanded. "Are you trying to put together a lawsuit against me?".

"I know I don't have the right to bring a lawsuit," the minister said, "but I preserve these so that the court historian will be able to record things accurately."

As soon as he heard that, Zhao Kuangyin reacted like the Monkey King when his master chanted the incantation to tighten the golden band around his head. He forgot all about hunting for birds.

Unbelievably, the court history really did record this incident. A hundred years later, Sima Guang (1019–1086), more famous for his *Comprehensive Mirror in Aid of Governance*, recorded the incident in his *Records of Rumors from Sushui*. He concluded the anecdote with a note that Zhao Kuangyin was both horrified and elated. He was horrified that the incident might be recorded, but elated to discover that his loyal minister's chief concern was accurate historical record. Zhaog Kuangyin calmed down and called for the minister to be gifted gold, silver, and silk. This was intended as a reward, and also to comfort him.

I will explain the *shiguan* system here. From ancient times, the daily life of the monarch had been recorded. The *Book of Rites* explained it like this: "Movements to the left; speech to the right." In other words, what the emperor did was recorded by an official historian that sat to his left, and what he said was recorded by an official historian that sat to his right. "The Emperor's actions must be recorded; good and ill, successes and failures; there can be no exceptions."[6] That means that every single thing that took place had to be written down.

All of these records were called *The Record of the Emperor's Daily Activities*.

From the time of Emperor Wu of Han, China officially began recording a *Private Record of the Emperor's Daily Activities*. In the Jin Dynasty, there came to be various secretaries in charge of recording various elements of the *Record*. When the emperor

---

[6] This comes from Xun Yue's *Shenjian*, or *Reaching For a Historical Mirror*, composed in the Eastern Han.

died, the *Record* would be passed to the *shiguan*, who would edit it into a coherent history and pass it on to the next emperor. It provided guidance to the next generation and also served as a memorial to the late emperor.

Looking at it from a theoretical point of view, the *Record* has a number of unique traits. The first trait is that it's true: it records true events, large and small, as they happened, without editorial or commentary. The second trait is that it cannot be changed, or, rather, that the cost of changing it is high. The third trait is that the monarch themselves were unable to look at it; letting the monarch see the *Record* was a violation of the ancestral system and would be opposed by all ministers of the court; keeping the *Record* from the monarch was a guarantee of its independence.

The existence of the *Record* was a major constraint on the monarch. One day, Emperor Taizu of Song returned to his palace and went to sit alone in his room, brooding. A eunuch named Tang Ji'en asked him the reason. Taizu said: "I was too hasty in the early morning session. I made errors. The historian recorded all of it. I have good reason to be upset, don't I?".[7]

Taizu's reaction shows us that although the authority to record was not power in the traditional sense, it still had the power to shape how the monarch spoke and acted. In the traditional bureaucratic system, the monarch had a heavenly mandate, placing him beyond earthly authority. That system meant that there were few restrictions on his authority. The *Record* not only changed the monarch's behavior, but could even change the trajectory of the feudal imperial system.

Although many monarchs expressed their dissatisfaction with this, it was all for naught. Think about the system that a monarch of that time would have lived under: "All land under heaven belongs to the emperor, and all people living on it are his subjects." Now, imagine that there is a method to record all of the monarch's actions and words, but he can't exercise authority over it… It had to have been an annoyance.

Even before Emperor Taizu of Song, Emperor Taizong of Tang had already expressed his dissatisfaction.

Li Shimin (598–649)—the Emperor's personal name—came to power through less than legal means. In 626, he stormed into the court and his troops began killing relatives of the Emperor. This is called the Xuanwu Gate Coup. Li Shimin took the throne and kicked off a golden age in Chinese history.

The great military man Li Shimin feared his deeds being recorded, too. The *Essence of Politics in the Zhenguan Reign* records that Li Shimin admitted to not wanting to speak often in court: "When I attend the morning session, I must weigh each word, considering whether I am speaking in the interest of the common man or not. I keep my words to a minimum."

---

[7] The source of this anecdote is the *Dushujing*, or *A Mirror for Reading*, composed in the Ming by Chen Xiru.

Du Zhenglun, who was in charge of the *Record*, exhorted: "My duty is to make sure that everything the Emperor says must be recorded in detail. If the Emperor says something that violates his imperial virtue, it will be known for thousands of years. It is good that the Emperor considers his words."[8]

Li Shimin always wanted to take a look at the *Record*. He wanted to know what was being said about his reign. One day he issued an imperial edict, demanding to see it. "I have no other goal but to examine it as a way to better understand my own errors." He must have thought that was the best way to sneak a peek.

"The Emperor has made no errors," Zhu Zishe, one of the high officials, responded. "It would not be a problem for you to look at the *Record*. But this would set a precedent for future historians. The *Record* is based on true actions and words. Whether those actions and words are good or bad, they were done and said by the Emperor. Some rulers have only wanted their good deeds recorded. If they were to look at the *Record* and see their failures noted down, they would be displeased. The official historian would fear for his life. If you were to look, it endangers future generations of historians, whose job it is to record the truth."[9]

When Emperor Taizong, he had no comeback. But he couldn't let it go. A short time later, he went to seek out Chu Suiliang, the minister in charge of admonition and arbitration. "You are in charge of the *Record* at present," the Emperor said, "so you must know what gets written in it. Can you let me look?".

Even though Emperor Taizong had taken power in a brutal coup, he still had respect for the institution of the official historian.

Unexpectedly, Chu Suiliang turned him down, too. "The man that is responsible for recording the *Record* is just the same as the two men that once sat at the left and right sides of the ruler, recording his actions and words. Everything has to be recorded—good and bad. The reason for this is that it is meant for rulers of future generations. It should provide lessons to future rulers on correct governance. I have never heard of any past rulers attempting to see the Record."

"If I make errors or speak inappropriately, will you record it?" Emperor Taizong asked.

"Of course," Chu Suiliang said. "It is my duty!"

Another minister named Liu Bo was standing nearby. "Even if Chu Suiliang does not record your errors," he said, "they will be recorded by the people of your kingdom."[10]

This was quite clever. It warned the Emperor off his attempts to burnish his image in the *Record*. He nodded his head and gave up on his quest to peek.

Li Ang (809–840)—the personal name of Emperor Wenzong of Tang—really did end up seeing the *Record*.

---

[8] The source of these lines is Wu Jing's *Essence of Politics in the Zhenguan Reign*, composed in the Tang.

[9] The source of these lines is the *New Book of Tang*, composed by Ouyang Xiu and others during the Song Dynasty.

[10] The source of these lines is Comprehensive Mirror in Aid of Governance, composed by Sima Guang in the Song Dynasty.

One day, Emperor Wenzong and his high officials were discussing political issues, while Zheng Lang stood nearby, making notes for the *Record*. After the ministers dispersed, Emperor Wenzong said: "Did you write down everything we just said? Let me have a look."

"What I have written is intended for the historical record," Zheng Lang said. "As you know, it is forbidden by convention for the Emperor to read it. Emperor Taizong wanted to read the *Record*, too. Zhu Zishe and Chu Suiliang had to dissuade him."

Emperor Wenzong was not impressed with that answer. "Anyways," he said, "the things we discussed today can't be easily divided into good or bad, so what does it matter? Let me have a look. My only intention is to review what I said, just in case I let slip any remarks that I should not have. If I could have a look, I would know to be more careful in the future."

Emperor Wenzong's remarks had their own logic. There had long been a debate about whether or not to reveal the *Record* to the ruler in charge. In most earlier dynasties, the rulers had been given the chance to see it. That practice had lasted up until the Northern and Southern dynasties, when the convention had shifted to the ruler not looking at the *Record*. From the Tang onward, the convention had been honored.

In the end, Zheng Lang could not hold the Emperor back. He let him see the *Record*.[11]

Not unsurprisingly, once Emperor Wenzong took a peek, he wanted to see it again. By that time, a new secretary was in charge of taking notes. His name was Wei Mo. He refused to let Emperor Wenzong peek. He advised the Emperor that if he was worried about how he was being remembered, that it was better to worry about doing good deeds. Emperor Wenzong was dissatisfied. "I have already looked at it once," he said, "so, why not let me look again?".

Digression: Should the ruler be able to look at historical records or not?

After the death of Emperor Taizu, he was succeeded by Emperor Taizong (939–997), who decreed that the *Record* should be sent to the official historian only after he had done a monthly review. Emperor Renzong (1010–1063) came to power, Ouyang Xiu criticized rulers that read the *Record*. He submitted a petition to the effect that the *Record* should remain independent. Emperor Renzong accepted his opinion. However, when Emperor Xiaozong took power, he broke with that convention. Hu Quan, a famous minister of the period wrote a petition requesting that he honor Renzong's example and not look at the *Record*.[12]

---

[11] The source of this anecdote is *Song Government Manuscript Compendium*, compiled by Xu Song in the Qing Dynasty.

[12] This is described in the *Book of Zhou*, composed by Linghu Defen in the Tang Dynasty.

The idea of rulers looking at these historical notes was supported by some. Northern Zhou Dynasty historian Liu Qiu (501–554) claimed that making everything public would be instructive to the ruler, his ministers, and everyone in the world. He wrote a petition to the ruler of the time: "Sovereigns of ancient times established the position of court historian not only to record events, but also to offer warnings. The historians that sat at either side of ancient sovereigns would praise their good deeds and rebuke poor decisions. This was ideal. … However, since the Han and the Wei, the *Record* has become a secret document. It is only to be passed onto future generations. That makes it useless to sovereigns about whom it is being written. This will not lead the sovereign to moral virtue and help them to avoid falling into bad habits. There is no way to know whether what is being recorded is true or not. This will lead to disputes and different ways to view the same facts. … Perhaps future generations reading the *Record* will also dispute their account. Nobody will know what is accurate. … I advise that the *Record* be written down and then make it available to read. This is the way to make things clear. By doing this, everyone can understand the successes and failures of the current sovereign. Let well-intentioned advice and benevolent actions be cultivated and erroneous actions shunned."[13]

"You were given a look," Wei Mo said, "because of a mistake on the part of the secretary. I will not neglect my duty. If we let you see the *Record*, we will not be able to do our jobs correctly. We might be hesitant to record your failures. Right and wrong would be confused. It would distort history. How could we bear to pass on such a *Record* to future generations?"[14]

Emperor Taizong and Emperor Taizu were both wise rulers. They respected convention enough to ask permission. There were many more monarchs far less reasonable than them. If they didn't like what the historian wrote, they would replace them. To curry favor with rulers, some historians would write elaborate odes to their virtues, neglecting to write a single word about any failures.

From this, you might be able to guess that being an official historian was a risky job in ancient China. A ruler could exercise his authority to make life pleasant for a historian that could ensure his posthumous acclaim. But a historian that wouldn't cooperate was risking the wrath of the ruler. There are many examples of historians cut down in their offices or their entire family lines being snuffed out. When Emperor Taiwu of the Northern Wei (408–452) saw the history recorded by Cui Hao, he believed that it exposed a plot against him, being led by the historian's clan. He could not tolerate dissent. He ordered the execution of Cui Hao's entire clan, killing at least 128 people.

---

[13] This is from the *Old Book of Tang*, compiled by Liu Xu in the Later Jin Dynasty.

[14] This is from the *Book of Jin*, compiled by Fang Xuanling in the Tang Dynasty.

Unfortunately, there were many foolish leaders in history. But even if they spent their reigns indulging in wanton luxury, they still hoped to leave behind a glorious record. If the *Record* they passed on was stained with misdeed, it would bring shame upon those that came after them. In 403 AD, Huan Xuan (369–404) usurped the throne of the Eastern Jin Dynasty and declared himself emperor. His rule was "arrogant and extravagant, distracted by hunting and leisure that seemed to go on day and night." He was later driven by Liu Yu (363–422) out of Nanjing. As he fled, his greatest concern was not staging a comeback or overcoming his difficulties, but what was recorded about him in the *Record*. He was dissatisfied with the historian's record, so he began composing his own. He wrote voluminously about his wise decisions and superb strategy. He argued that it was only because his subordinates had failed to heed his commands that he had been driven from power. Once he was finished, he issued a "proclamation far and wide," hoping to popularize his own account. Dictating the text and having it dictated back to him for editing took up most of his time. There was no chance to convene meetings of his generals. Within months, his troops were defeated and he was beheaded.[15]

Good rulers fear recording, but so do lesser rulers. Knowledge, philosophy, and prestige come from observing, analyzing, and summarizing history. Historical records represent objective reality. History is perpetual and transcendent. There is a line drawn from traditional philosophy to the effect that the laws of nature would continue to obtain, whether a benevolent sage-king such as Yao reigned or whether a wicked tyrant like Jie reigned—and this can be applied, also, to history. In a society that was traditionally based on Confucianism, the conduct of a ruler was closely linked to their legitimacy. The ethics of virtue and morality were paramount, so the ruler had to do their utmost to uphold them. When it came to their speech and behavior being recorded, they had no solution but to perform for history.

Recording can inspire our internal human ethics. We still have not come up with a good way to explain the psychological secrets behind those ethics. Kant put it this way:

> Two things fill the mind with ever new and increasing admiration and awe, the more often and steadily we reflect upon them: the starry heavens above me and the moral law within me.

Basically, recording has a mysterious power. In order to counter this force, Confucians came up with the idea of *shendu*, which means to be absolutely blameless in your personal life. If we can hold ourselves to that, there is no need to fear having our words or actions recorded. Human history moves in cycles. In recent years, some rulers were not afraid of their deeds being recorded, but of not being recorded.

I am referring to the age of the democratically-elected president. We have to return to Edison because it was him that expanded our recording methods beyond writing, painting, and photography.

---

[15] This is from Sima Guang's *Comprehensive Mirror in Aid of Governance*.

5.3   Edison Broadens the Horizons of Recording

In 1877, Edison was experimenting with his telegraph and made a discovery: mechanical vibrations could produce a sound. He came up with the brilliant idea that mechanical vibrations could be generated in a way that reproduced a sound. Edison and his assistants worked around the clock to develop the first phonograph. Edison recorded himself reciting "Mary Had a Little Lamb." He played it back from a mechanical device. It was like an extraterrestrial transmission. People that heard it were dumbfounded. It was a watershed moment in a new age of voice recording.

Edison later explained the principle and the process in an article:

> We have all been struck by the precision with which even the faintest sea-waves impress upon the surface of a beach the fine, sinuous line which is formed by the rippling edge of their advance. Almost as familiar is the fact that grains of sand sprinkled on a smooth surface of glass or wood, one or near a piano, sift themselves into various lines and curves according to the vibrations of the melody played on piano-keys. [...] This led me to try fitting a diaphragm to the machine, which would receive the vibrations or sound-waves made by my voice when I talked to it, and register these vibrations upon an impressible material placed on the cylinder.

Edison's life gave us many monumental inventions, but the phonograph caused the biggest sensation (Fig. 5.1). Forty years before, there had been amazement that photos could reproduce events from the past, and there was amazement again when the phonograph became a "speaking machine."

There are two main ways that we absorb information: our eyes accept visual information and our ears accept auditory information. Writing and images provided a way to record visual information, and Edison's invention provided a way to record auditory information. The phonograph moved us from only recording visual data to also recording auditory data. For the first time in history, the phonograph gave us a way to record invisible, intangible, fleeting sounds in a way that allowed for playback at a selected time.

Now, we are in an age of AI. The core of AI technology is an ability to imitate the human brain's ability to process visual and auditory data and make decisions about it. We want AI to be able to read the written word, make out the details in pictures, and be able to understand speech. Those abilities are what will allow them to become closer to ourselves and confirm their intelligence. Having said that, it reminds us that Edison's speaking machine was the first artificial intelligence.

Edison's phonograph was sent all over the world, even ending up in the hands of Kaiser Wilhelm II (1859–1941). The Kaiser loved his phonograph to the point that he gave his ministers an impromptu lecture on the principles behind it:

> They listened patiently as he expounded on acoustics, sound waves, and mechanical vibrations, but they were astounded when he inserted a cylinder, fired up the motor, and his disembodied voice filled the room. The excitement was hard to contain. The lecture and demonstration continued for several more hours. The Kaiser explained the working details of the phonograph in such detail that he could have been mistaken for its inventor. Once he was done, he left the room, leaving his mystified ministers behind to puzzle out everything they had just seen and heard.[16]

---

[16] Jones (2016).

Fig. 5.1   Edison shows a tinfoil phonograph in 1878[17]

Reading the history of the phonograph, I was amazed by the reception of technological innovations among the European aristocracy of the time. Emperor Kangxi of the Qing had a fondness for clocks and watches, but the spirit of science did not take root in his dynasty. Technology and new inventions were seen as being on the same level as magic tricks.

The first Chinese person to encounter the phonograph was Guo Songtao (1818–1891), Ambassador to the United Kingdom. In May 1878, he was at a social event where Edison was demonstrating his invention. Guo Songtao described it in his diary as "heavenly technological masterpiece." The phonograph was introduced into China through Tibet in 1897. Merchants there knew that there was a need for repetitive prayer recitals. Recording prayers would be a good money making opportunity. They went and recorded a lama reciting a Sanskrit mantra, "Om mani padme hum." The recording was played back on a loop. The lamas were stunned.

The phonograph surprised crowds in London, too. One day, a large crowd gathered around a hawker's cart. Under normal circumstances, a hawker's cries wouldn't cause much of a stir—but on that day, they were coming from a phonograph. As the scene unfolded, the hawker himself stayed mostly silent, taking cash and handing goods

---

[17] This photograph by Mathew Brady is from Oxford English Dictionary Online.

to the customers. It turned out that the hawker had come down with the flu and lost his voice. When Edison heard this story, he burst out laughing.

Digression: the difference between the tape recorder and the phonograph

Edison's invention was called the phonograph; in April of 1878, he founded the Edison Speaking Phonograph Company. The playback ability of the phonograph was crucial. As sound vibrated a piece of metal, a stylus carved grooves into a piece of tin foil. When a stylus was run back over the grooves, it reproduced the sound. This invention laid the foundation for the tape recorder. Different designs for the tape recorder were invented successively in Germany and then the United States in 1938. The principle they worked on was the same: a magnetizable medium was run past a recording head. This became the twentieth century's most popular recording system.

Shortly after inventing the phonograph, Edison predicted that it's primary utility would be in the office. His idea was that the phonograph would be used to record conversations and phone calls for clerks to transcribe. It could also be used in the courtroom, replacing the stenographer in recording remarks by the judge, lawyers, and witnesses. He also guessed that it would be used to record music.

Edison was correct. The later development of recording technology proved his foresight. In 1940, the White House received a tape recorder and President Franklin D. Roosevelt ordered it installed in the Oval Office. Edison predicted a bright future for his invention, but nobody at that moment could have predicted that it would change the course of American political history.

5.4   Nixon's Dilemma: A Brief History of White House Sound Recording

The 32nd President of the United States, Franklin Roosevelt wanted to introduce recording equipment into the White House because of a strained relationship with the press.

Roosevelt often complained about reporters taking his remarks out of context or blatantly misquoting him. He had spent a lot of time fighting wars of words with the media. In February 1939, he called a press conference to criticize reports:

I have in front of me, oh, about eight or ten different newspapers. There isn't one story or one headline in all of those papers that does not give, to put it politely, an erroneous impression-not one. It is a rather interesting fact. These things have been manufactured by deliberate misrepresentation of facts, existing facts.[18]

---

[18] Bennetts (1982).

When Roosevelt was seeking a third term in 1940, with the Second World War raging. At that time, the American public was isolationist and Roosevelt's pro-intervention remarks often caused an uproar. Wendell Willkie, Roosevelt's Republican challenger attacked him for this, charging that Roosevelt was going to send American sons, brothers, and loved ones to their graves. Roosevelt had to be very careful with how he spoke on foreign policy and wartime strategy. He once said at a press conference that he did not want war with either Hitler or Mussolini.[19] When this statement was reported in the papers, it was taken as a lack of stamina for combat. For Roosevelt, explaining himself to the press took up more effort than it was worth.

For this reason, Roosevelt liked to bypass the newspaper and radio reporters and go directly to the public through his fireside chats. He also befriended NBC's David Sarnoff, who understood Roosevelt's difficulties with the press, and was the one that gifted the White House a tape recorder. Roosevelt had it installed under a lamp on a desk in the Oval Office. Whenever he had to talk to the press, his stenographer would flip the switch on the tape recorder.

Roosevelt warned reporters not to misquote him, since he had the recordings as backup. Altogether, there survive fourteen recordings of press conferences, but for reasons that remain unclear, there were recordings made of several private conversations. In these unguarded moments, we can hear Roosevelt talking rather coarsely about Willkie's rumored extramarital affairs.

Following Roosevelt, Truman, Eisenhower, Kennedy, and Johnson all used tape recorders in the White House. Kennedy was particularly enthusiastic about recording conversations in the Oval Office, especially after the media roasted him over the disastrous Bay of Pigs invasion.[20] By the time Johnson replaced Kennedy, recording technology had progressed to the point that it could easily be installed in more White House offices, as well as patched into phone lines, allowing every meeting and call to be recorded.

At that time, the White House recording system was top secret and its existence was known to very few. 1973 and Richard Nixon would change all that. Watergate shocked the world and wrecked Nixon's second term—and it also made public the White House recording system.

Nixon entered the White House on January 20th, 1969. After the inauguration ceremony, he and Chief of Staff HR Haldeman (1926–1993) went into the Oval Office and opened the pantry over the fireplace, which Johnson had told them about. That was where the recording equipment was concealed. Inside the pantry was a control panel, connected with wires to other offices in the building. The two political veterans were shocked. They guessed correctly and were a bit frightened by the idea that Johnson must have recorded many of the conversations held at the White House.

As president-elect, Nixon had been informed that there was a recording system, but he had no idea of the extent of its use. Johnson had once complained to him

---

[19] More specifically, he said: "We don't want any war with you. ... We don't consider ourselves belligerent."

[20] This refers to a failed April, 1961 invasion of Cuba by Cuban refugees assisted by the CIA. The operation was a failure.

that everyone that left his office had their own ideas about what he was thinking or plotting. For their own benefit, Johnson said, people would distort what he had said. The recordings were his way to keep the facts straight.

This had been a problem for previous presidents, too. Eisenhower had admitted in 1954 that he recorded, saying that it was the best way to protect himself from people in Washington that he didn't trust.[21]

Two people have a conversation and, afterwards, one of them attributes something he said himself to the other person. There's a slang term for that: "flipping the pancake."

English philosopher Francis Bacon once wrote an essay on "flipping the pancake." He tells the story of two friends that are vying for a ministerial appointment. One says that he would rather not be a minister, since the monarchy is in decline. The second man agreed. Later, when the first man reported the conversation to others, he told them that his friend was uninterested in becoming a minister. When the Queen heard the remarks, the first man received the ministerial appointment and his friend fell out of favor. He "flipped the pancake" on his friend, baiting him into agreeing with him but then only reporting one side of the conversation.

There are many examples throughout history of this trick being played. The reason is simple: nobody had the ability to record conversations.

Nixon understood this. He had served as vice president for two terms before taking the Oval Office. But he wasn't fond of secret recordings. He thought experience and diplomacy would carry him through any complicated situations. The very day he and Haldeman discovered the system, he ordered it removed.

Nixon quickly encountered difficulties.

First, there was no satisfying way to take minutes in a meeting. Nixon's staff had arranged for a stenographer but there were too many meetings for them to keep up. It became inconvenient for many people to talk at once. Nixon was not pleased with the situation. Apart from the inconvenience of the stenographer, he found that the written word could not catch certain unspoken but important pieces of information, like a subtle change in a speaker's tone, for example. Haldeman made a suggestion to use different color codes to mark tone or emphasis, but this was somewhat limited.

What Nixon noticed about his notes shows some of the limitations of using the written word to record. It's a system of one-dimensional symbols, unable to capture all the complexities of reality. A person's tone, expression, gestures, and other body language all add to what a speaker is saying, but a transcript has no way to record this specific information and how it transpires in time and space.

Second, it was hard to attend to diplomatic requirements. Nixon played the diplomacy card during his first term, so he was often in meetings with foreign dignitaries. In order to build mutual trust and warmth, he would often rely on his guest's interpreter, instead of bringing in one of his own. During the talks with China in 1972, Nixon relied completely on interpreters from the Chinese side. There are obvious disadvantages to this practice, however. A small mistake in the interpretation can

---

[21] Ting (2001).

lead to immense confusion, which can affect policy decisions. Nixon needed recordings of his meetings so that an interpreter from the White House National Security Council could review the tapes, checking for inconsistencies.

The final reason for the recording was more selfish. After American leaders leave office, they go back to being private citizens. They step back into the shadows. Most presidents compose memoirs as a way to pay the bills. With all the recordings, Nixon knew that his book could be as accurate and complete as possible.

At the beginning of 1971, Nixon reversed his earlier decision and had recording equipment reinstalled in the White House. He made one request of Haldeman: he wanted a single switch to toggle the system on or off, so that he could always have it ready. "Mr. President," Haldeman said, "I can promise you that you'll forget to hit that button. By the time you realize it, it'll be too late. I know what's going to happen."

Haldeman identified a common dilemma that many of us have faced: we find ourselves unprepared in situations we should be prepared for. He had a solution, however. Technology had improved since the Johnson administration. It was possible to have the system be sound-activated. As soon as someone entered the office or flicked on the light, the tapes would start rolling.

Nixon approved this high-tech solution. He never expected that this decision would one day lead him into an abyss.

The recording system began operating on February 16th, 1971. The system was restricted at first to the Oval Office and the Cabinet Office, but Nixon had his deputy assistant Alexander Butterfield expand it out to other offices, meeting rooms, and even living areas. In May of 1972, Camp David was outfitted with its own system. Just as Haldeman had promised, as soon as anyone entered a room, the mics would be hot and the tapes rolling.

Of course, the existence of this system was highly classified. Nixon had repeatedly stressed that the tapes were for his personal use only. Apart from the technicians that installed the system, only Nixon and the three men closest to him were aware of its existence. The entire Cabinet, including Secretary of State Henry Kissinger were completely unaware that they were being bugged by the President.

After installing the system, Haldeman had it tested several times by technicians to make sure it was working normally. It was quickly forgotten as the White House began normal operations. Nobody had time to pay attention. "[T]he worry quickly passed," Haldeman recalled, "as did any awareness at all that Nixon's conversations—my conversations, often—were being taped. I think Nixon lost his awareness of the system even more quickly than I did." He only occasionally considered that he was being recorded: "I sometimes ask myself if I would have said some things differently if I had consciously considered the fact that my words were being taped."[22]

Of course he did. "But my confidence," Haldeman said, "that the tapes were never going to be heard by anyone except Nixon and myself was so great that I really do suspect I would have pushed any incipient worry about disclosure aside and spoken just as I did."

---

[22] Haldeman (1988).

Although the system was filling up tapes, Haldeman recalls Nixon rarely listening to them. The only significant incident was near the beginning of 1972, when Henry Kissinger took an interview with an Italian journalist and seemed to be trying to distance himself from Nixon's Vietnam policy. Nixon was furious and told Haldeman to warn Kissinger that all of his conversations were being recorded and to stay on message.

Haldeman, however, didn't follow this command. He secretly modified the order: "I am virtually certain I never gave Kissinger this message. I chose to consider what Nixon said to me more as a flare of anger than as considered instruction, and I probably told Kissinger simply to be more careful about what he said during interviews."

Did this kind of large-scale monitoring program have its risks? Nixon himself was not so sure it should continue. On April 9th, 1973, he ordered Haldeman to dismantle the system. By the afternoon, he had changed his mind. Nixon wanted the system maintained but an on–off switch installed. Haldeman was distracted and neglected the order. The sound-activated switch remained in place. The tapes were always ready to roll.

When the Senate Watergate Committee began looking into the scandal, Butterfield was hauled in to testify. On July 16th, 1973, he was asked if any listening devices had been installed in the White House; he was forced to answer in the affirmative, explaining that all conversations in the White House had been recorded. This was a bombshell. It seemed to be proof that the White House held the key to breaking open the investigation. The Senate Committee immediately subpoenaed the tapes.

Two days later, the tapes finally stopped rolling at the White House.

In fact, even before Watergate had become a scandal that publicly implicated himself, Nixon had started to be uneasy about the recordings. In April of 1973, he had ordered Haldeman to compile all recordings of a March 21st, 1973 conversation with White House aide John Dean about Watergate. "He wanted to know precisely what he had said during that troubling conversation," Haldeman recalled.

Nixon had refused to give up the tapes to the Senate on the grounds of executive privilege. The matter went to the Supreme Court, who, after three weeks of deliberation, ruled unanimously (one Justice recused himself due to connections to the Nixon administration) that Nixon was bound by the law to release the tapes to investigators.

Under immense pressure from the Supreme Court and political rivals, Nixon relented and handed over the tapes.

Compared to the eight hours of recordings left behind by Roosevelt, the 4,000 h Nixon tapes represent an immense amount of data. When these recordings were finally transcribed, they ran to 27,000 pages. 12.5 h of recordings were singled out as key evidence in the Watergate investigation. These recordings included the so-called "smoking gun" tape, which covers the period six days after the arrest of the operative that broke into the Democratic Party offices in the Watergate Hotel to install a wiretap, during which Nixon began planning a coverup.

The final result was that Nixon was forced to resign, becoming the first American president to step down over impeachment. Everyone else involved in Watergate received varying degrees of punishment. In fact, the tapes that Nixon had submitted to the Supreme Court were incomplete. There was a gap in the recording that lasted

18.5 min, supposedly the result of an error by Nixon's secretary. The contents of the gap remain a mystery to this day, with some suggesting that it contains Nixon and Haldeman talking about specific arrangements for the wiretapping of the Democratic Party offices. If that was true, it would have been enough to send Nixon to prison.

Nixon denied these allegations, of course. Following his resignation, he spent his remaining years trying to restore his tattered reputation. In memoirs he published in 1990, he described his original intentions for his recording system.

From the beginning, he intended for his administration to be "the best chronicled in history." He "wanted a record of every major meeting" that would range "from verbatim transcripts of important national security sessions to 'color reports' of ceremonial events."[23]

5.5 Trump's Recording Troubles

In the intervening decades, the White House recording system faded in memory—until one day in May of 2017, when it hit the headlines again. Nixon was replaced as the protagonist in this story by the 45th President, Donald J. Trump.

As soon as Trump was elected, "Russiagate" flared up again. This was centered on a suspicion that Russia had been involved in or influenced the American general election. Figuring out what precisely happened became a matter of Congressional and FBI investigation. Shortly after taking office, Trump had invited FBI director James Comey for lunch. On May 9th, 2017, Trump abruptly relieved Comey of his duties, far ahead of the end of his tenure.

Soon afterward, Comey went to the New York Times to reveal what had been discussed at lunch that day. Trump, Comey said, had asked him about the progress in the Russiagate investigation and requested his loyalty. He also asked for leniency for former White House aide Michael Flynn. As Comey told it, he had expressed to Trump that he would justify the law impartially; in the end, the promise to uphold the law had caused Trump to relieve him of his duties and seek a replacement. Trump promptly denied all of Comey's one-sided claims. The same issue arises again: if two people are giving their account of a private conversation, neither is reliable.

Three days later, Trump took to Twitter to attack Comey, posting a warning that he might have a recording of the conversation. The implications were clear: if there was a tape of the lunch chat, Comey had best watch what he says.

This touched off a heated discussion in the media and in public opinion. Forty years after Nixon, a president was hinting at a system to record conversations.

After Watergate made public the White House recording system, there had been a debate in the United States over who had the right to record in the president's office and who should own the recordings. In 1978, Congress introduced the Presidential Records Act to govern the president's official records. It had three key points: public ownership of presidential records, a statutory structure for presidents and

---

[23] Nixon (1990).

vice-presidents to supervise and manage their records, and rules must be followed about differentiating official and private recordings.

Nixon's disastrous experience had been enough to put future presidents off recording in their White House. That changed with the 40th president, Ronald Reagan. He used it occasionally in the belief that recordings could confirm the accuracy of memory. He went beyond simple audio recordings to capturing video of meetings with foreign guests. Reagan was accustomed to being in front of the camera, having been a Hollywood star before entering politics. When Reagan biographer Craig Shirley was asked whether the former president would have ever used the tapes improperly, he said, "Threaten or attempt to blackmail a former government official? The question itself is preposterous because it never would have even crossed Reagan's mind."[24]

Trump's Twitter post to Comey about recordings was clearly a threat. There are many examples throughout history of people attempting to use records in their possession to bully opponents. It's possible that Chinese history has the finest examples of this practice. Ren Bo'an held a position in the Qing bureaucracy akin to today's office clerks, filling his time drafting basic official documents. He used his position to begin noting down failures, secrets, and improprieties of various types. The records he collected became a powerful blackmail tool. Nobody had the nerve to turn down anyone wielding those records. The nobles of the Yongzheng did everything they could to recover them. It became a political scandal.

Trump's threat worked and Comey went quiet. But that wasn't the end of his troubles. Given his lack of political experience, it's likely that Trump didn't understand the Presidential Records Act. Within days of making the threat, the Senate ordered Trump to hand over any recordings he had made of himself and Comey. In accordance with the 1978 law, any recordings made by Trump belonged to the state. That would mean that anything he recorded would be reviewed and then archived, with the possible option of making it public.

On June 22nd, Trump fired off two more social media missives, denying that he had any recordings. He said, essentially, that he was aware of conversations about withholding electronic surveillance, but, if any recordings existed, he had not made them and didn't possess them.

Trump was exploiting a loophole in the 1978 Act. According to the law, any recordings made by the White House belong to the state, but that does not prohibit the President from making recordings as a private citizen. With the development of technology, recording equipment does not require a system like Nixon installed in the White House. A high quality miniature recorder could fit in your pocket and be triggered at any time. It could also be done with a phone. The 1978 law didn't specifically ban any of this, so it became something of a legal gray area.

The American press speculated that Presidents must have used recording equipment in the past. Even though Obama never made use of the White House system, he was still making tapes. When Mark Bowden was interviewing Obama for a book on the killing of Osama bin Laden, he suddenly panicked when he realized his miniature

---

[24] Montanaro (2017).

recorder was not working, but once the interview was over, the Obama team passed him a complete transcript. That was proof that Obama was making recordings of interviews and meetings.

Trump's situation was not that different from that faced by Roosevelt. He didn't trust the news media and publicly criticized them for distorting facts. His accusation of "fake news" turned the phrase into a buzz word in the United States. While Roosevelt had taken to the radio to get his message out without distortion, Trump went to Twitter. During his first year in office, he made an average of six to seven posts on Twitter everyday. Perhaps this "Twitter governance" will one day be studied by future generations.

Whether broadcast on the radio or posted to Twitter, the information is recorded. Over the past century, our capacity and ability to record has continuously expanded, starting with text, then images, then sound, and video. In a political environment always in flux, the written word is too vague and images have their own limitations. The accuracy of audio records is much greater, and they have the benefit of being able to be produced much more discreetly. Holding on to records is ostensibly an aid for memory and to improve accuracy, but it also means possessing secrets, truths, and evidence. What Roosevelt set in motion in the White House has not yet reached its conclusion.

5.6 The Inefficiencies of Live Broadcast

As we've already seen, when Reagan entered the White House in 1981, he began using video in addition to audio. Video technology was becoming more advanced and popular. In 1979, the American government announced that it would begin live broadcasts of sessions of Congress and committee debates.

The Constitution stipulated that all votes in each House should be recorded. When the United States Congress was first convened in 1789, a staff was established to record meetings and debates of the Senate and the House of Representatives. Orders were made to disclose all of these records to the public. At that time, there was no Internet, or even television, so the information was mostly dispersed through newspapers.

Beginning in 1841, Congress had a press box, but, like Roosevelt and Trump, they still had problems. They accused reporters of changing quotes, misleading readers, or even creating "fake news." There were numerous disputes that led to reporters occasionally being expelled. As the two sides battled, it was the public that suffered from lack of information.

Kennedy's presidency saw the rise of television. In 1979, Cable-Satellite Public Affairs Network (C-SPAN) was established to live broadcast Congress. C-SPAN had an outsized influence in the political culture of the United States. Brian Lamb (1941-) , C-SPAN's founder said that TV could reveal the entire political process and give citizens the feeling of being present for deliberations. With a complete broadcast record of goings-on, there were far fewer disputes between members and the press.

In April of 2012, Congress announced that they would live stream coverage at HouseLive.gov and make archival recordings available. Meetings often last late into the evening and can run more than ten hours. But all of it was available online, from the opening prayer, to the speeches by members, to the debate over votes, to voting, to counting the votes, and even the adjournment process. Every word and action would be broadcast and recorded. That was expanded from regular House meetings to less formal events with individual Congressmen and their staff. Along with the video, the website also hosts additional material about the meetings. Users can quickly search for clips or documents with keywords.

This is a comprehensive and continuous recording. Compared to *The Record of the Emperor's Daily Activities*, it is very impressive. Legislators know that every word they say, every gesture, and all their body language will be broadcast and archived for countless viewers at present and in the future. The politicians are aware that the voters will make a judgment on them. TV became a key place for politicians to put on a show and attempt to draw votes. A survey by the House showed that live broadcast of meetings had increased the number and length of speeches. Some members will even intentionally court controversy, since they know it will be recorded and make news. Since the time Congress passed their bill to broadcast all meetings, average speech time has increased by 54% and daily session time has increased by 6%.[25]

The paradox is that although recordings are meant to ensure truthfulness and accuracy, it can also distort human behavior.

In 2002, the 16th National Congress of the Communist Party of China proposed the idea of political civilization, but there have been few discussions over the past fifteen years of how to actually build it. We have already talked about the significance of records to what we could call Commercial Civilization and the same should be applied to politics. Accurate records of political goings-on could be a concrete measure to build political civilization. The United States has recorded every meeting of its Congress, meaning that two centuries of political debate have become an invaluable resource in political big data. We need to keep in mind that there is nothing new under the sun: for any problem, a solution has already been undertaken or proposed, so the idea of being able to search through centuries of political debate is very valuable. This immediate access to records is incredibly convenient to future generations trying to analyze and solve political impasses. Civilization comes from accumulation. We should stand on the shoulders of previous generations. These previous generations include all of humankind. Regarding the American Congressional records, there's no reason they shouldn't be translated into Chinese and referred to for our own political decision making.

---

[25] This is from a survey by John Anderson, included in Frantzich, S. "Communications and Congress" (1982). *The Academy of Political Science.*

5.7 Universal Recording: God's Ultimate Weapon

This section attempts to make a bold and hopefully interesting assertion.

It begins with a novel: *Adam Bede* by George Eliot (the pen name of Mary Ann Evans, 1819–1880). In the nineteenth century, she was just as famous as Charles Dickens or the Brontë sisters. *Adam Bede* tells us an emotionally tortuous love story.

The story takes place in the nineteenth century and starts with the heroine, Hetty, a beautiful country girl, being courted by the titular Adam, a local carpenter. Although she admits he would make a great husband, she allows herself to be seduced by the young squire Arthur Donnithorne. Adam and Arthur were originally friends, but when the former comes across the latter attempting a tryst with Hetty, they end up fighting a duel.

Arthur feels ashamed and ends up breaking up with Hetty in a letter, after which he leaves town to join a militia. Out of desperation, Hetty agrees to marry Adam, but then quickly realizes that she is pregnant. She runs off to find Arthur but ends up giving birth before finding him. After deliberating for a while, she abandons the infant.

Hetty is arrested and tried for the crime but refuses to plead guilty.

I don't necessarily want to talk about the love story but how Eliot constructs the narrative to show Hetty being awakened to her crimes and finally confessing and repenting.

Before she is set to be hanged, Hetty speaks with Dinah, her cousin:

> "But, Hetty, there is some one else in this cell besides me, some one close to you." Hetty said, in a frightened whisper, "Who?"
>
> "Some one who has been with you through all your hours of sin and trouble—who has known every thought you have had—has seen where you went, where you lay down and rose up again, and all the deeds you have tried to hide in darkness. And on Monday, when I can't follow you,—when my arms can't reach you,—when death has parted us,—he who is with us now, and who knows all, will be with you then. It makes no difference—whether we live or die, we are in the presence of God."

Of course, she's talking about the Christian God. More than anything said by the judge or lawyer, this causes Hetty to waver:

> Hetty obeyed Dinah's movement, and sank on her knees. They still held each other's hands, and there was long silence. Then Dinah said—".
>
> Hetty, we are before God: he is waiting for you to tell the truth."

Dinah prays with her:

> Fear and trembling have taken ahold on her; but she trembles only at the pain and death of the body: breathe upon her thy life giving Spirit, and put a new fear within her—the fear of her sin. Make her dread to keep the accursed thing within her soul: make her feel the presence of the living God, who beholds all the past, to whom the darkness is as noonday; who is waiting now, at the eleventh hour, for her to turn to him, and confess her sin, and cry for mercy—now, before the night of death comes, and the moment of pardon is for ever fled, like yesterday that returneth not.

After a long time, Hetty cries out and embraces Dinah. She says that is willing to confess everything:

"Dinah," Hetty sobbed out, throwing her arms round Dinah's neck, "I will speak... I will tell... I won't hide it anymore. ... I did do it, Dinah... I buried it in the wood... the little baby... and it cried... I heard it cry... ever such a way off... all night... and I went back because it cried."

She paused, and then spoke hurriedly in a louder, pleading tone.

"But I thought perhaps it wouldn't die—there might somebody find it. I didn't kill it—I didn't kill it myself. I put it down there and covered it up, and when I came back it was gone."

What caused Hetty to confess was not the law or her conscience but the eyes of God. Dinah's description of Him as omniscient and ever-present made Hetty afraid. The idea that he knew everything that she said and did made her confess.

The dialogue that Eliot writes between them and the structure of the plot reflects the culture and customs of her own society. It shows us the most fundamental understanding of religion, which is that words and deeds are always being recorded in Heaven. God or his messengers see every sin and then record it with complete accuracy in a dossier. Nobody can hide from or deny their crimes. They can't deceive god. Facing an omnipotent and omniscient God, Hetty knew she had to admit the facts. Western religions tend to take this idea of an ever-present God to the extreme.

There are many examples from the Bible:

The sin of Judah is written with a pen of iron, and with the point of a diamond: it is graven upon the table of their heart, and upon the horns of your altars; whilst their children remember their altars and their groves by the green trees upon the high hills (Jeremiah 17:1–2).

Oh that my words were now written! oh that they were printed in a book! That they were graven with an iron pen and lead in the rock for ever! (Job 19:23–24).

Then they that feared the Lord spake often one to another: and the Lord hearkened, and heard it, and a book of remembrance was written before him for them that feared the Lord, and that thought upon his name (Malachi 3:16).

Notwithstanding in this rejoice not, that the spirits are subject unto you; but rather rejoice, because your names are written in heaven (Luke 10:20).

Christianity talks about original sin, repentance, and judgment by the Lord at the end of days; Chinese Buddhism talks about karma and six great divisions in the wheel of karma. The Buddhist sutras say that fate will catch up with everyone in the end. In other words, good is repaid with good and evil is repaid with evil.

Karma is still administered by a heavenly record, though. I would like to excerpt part of a sermon that explains this:

The Buddha sees us very clearly. We can't see him, but he sees us—why is that? He is always with us. He perceives the world but is not confused by it. We are confused by perceiving the world. He sees us. We can't see him. He knows our every movement. He is so great that he not only knows our movements but knows them better than us. The Bodhisattvas are the same, as are all spiritual beings. When the King of Hell looks you up in his Book of Life and Death, he will see if you have done more good than evil, and then divide those good deeds according to the three categories; if you have done more evil than good, those evil deeds will be divided according to three categories, and you will be reborn into this world.

Nothing can be falsified. Can we come to our senses, repent, and change our ways? Yes, we can return from evil to good. We can slowly begin to outweigh or eliminate the bad recorded in the King of Hell's book with good deeds. Our goodness will slowly grow and increase, but we are still unable to escape the six great divisions in the wheel of karma.[26]

This is essentially the same idea as Eliot was expressing in *Adam Bede*. The Buddhas and Bodhisattvas are watching us attentively and recording our word and deeds, so we had better be good. This is also the logic of Christianity.

In human history, the power of religion to enlighten and guide people was obvious. In religious doctrine, the power of a record is a deterrent force applied to everyone. God is watching everyone. What we do in this life will be recorded to determine our treatment in the next. If you do bad, God will punish you; if you do good deeds, the Buddha will smile on you. In other words, in order to deter adherents from bad and put them on the right path, Christianity and Buddhism both arrived at the concept of spiritual beings that record the thoughts, words, and deeds of all sentient beings.

The power of religion rests in the ability to record. That is the core mechanism of religious systems: recording things and meting out punishment or reward.

In Chinese folk beliefs, this is made even more concrete in the idea of the Book of Life and Death. Everyone has heard of the King of Hell, who is usually depicted with the book in his right hand. Legend has it that he has records going back 99 generations. The book is a record and the King of Hell makes judgements about the afterlife or reincarnation based on it.

The metaphorical Book of Life and Death is embedded in Chinese culture. In the first scene of the modern Beijing opera Guerillas on the Plains, the military hero of the story proclaims to some fighters from the puppet regime:

I'm telling you, the fighters and volunteers in the War to Resist Japan hold the Book of Life and Death in their hands. If you do not aid the cause, you can earn a black mark beside your name. If you stay on the road of righteousness and do good deeds, in addition to the medals you earn, you will have a red mark beside your name.

One of the soldiers he's addressing immediately calls back, "We want red marks!".

This analogy of black and red marks clearly outlines the form and structure of the Book of Life and Death as a record.

Going back to the beginning of this chapter, the ancients invented the The Record of the Emperor's Daily Activities and history books as a way to use the power of records to restrict the power of the monarch and the elite. In 709, after reading a history by Zhu Jingze, Tang minister Wei Anshi sighed and said: "Most people don't realize that the historians have more power than I ever will. I can only control the living; historians control the living and the dead. That is why some ancient emperors were so afraid of their historians."[27] This is proof that even at the earliest appearance of recorded data, its power was not overlooked. This insight led to the religious record idea being widely propagated. In modern times, we have still and

---

[26] Master Chin Kung. *2014: Notes on Pure Land Scriptures* (September 9, 2015). Quotations from a speech delivered to the Hong Kong Buddhist Education Foundation.

[27] This is adapted from the *New Book of Tang*, compiled in the Song dynasty by a group of scholars that included Ouyang Xiu and Song Qi.

video cameras, tape recorders, mobile phones, and advanced text records. Compared to the historical records of imperial China, our data is not only accurate but also structured, machine-readable, easily transmissible, and highly durable. Recording technology once possessed by a minority is now available to everyone. People don't only point these devices at each other, but use their technology to record the heavens, the earth, and everything in between.

As I said in a previous chapter, the basic scientific method starts with recording and measuring. Measuring is quantifying what has been recorded. The results of recording and measuring are data. Whether we are talking about an astronomer looking at a star or a social scientist looking at a person, the key to recording is establishing a trajectory—the path an object or person takes in time and space. But when it comes to people, they will not only chart a course through time and space but say and do different things along the way; they will have different moods, or different body language. To study a person, a complete record must be made.

Today, we are reaching a comprehensive record. Emperor Taizu of Song and Emperor Taizong of Tang should count their blessings: they didn't have to face modern technology but could attempt to directly influence the historians. All of their words and actions might now be recorded from multiple angles, with multiple devices and methods. These recordings could all corroborate or reinforce each other, meaning that no single person or institution had a monopoly on the truth.

The human lifespan can be a century. If we trained a camera on a person to record their daily life, there would be about 4 gigabytes (GB) of data generated per day, which adds up to about 143 terabytes (TB) over a hundred years. Going by present day prices, it would cost about 50,000 yuan to store those 143 TB; if the data was compressed correctly, the price could be reduced to about 20,000 yuan. That's how much it would cost to record a person's entire life.

The day is fast approaching that this will be a reality. As more people strap cameras to their heads and point their cameras at their faces, the age of complete lifetime recordings will come about sooner than we expected. Can we compare this power of complete, uninterrupted recording to that of God?

I contend we can, but our ability to maintain universal recording will one day surpass even God. There is only one God, but the recording tentacles that penetrate into our daily lives are innumerable. God does not have the capacity to track the lives of billions, but universal, round-the-clock recording technology will have no such limitations.

God will no longer be a fiction but a reality. Universal recording is a lens on human nature. Under this lens, human behavior will be gradually exorcized of its demons. We will more clearly see the good and evil, the trivial complaints, and the difficulties in human existence. The power of data will allow new social formations; lifestyles will be revolutionized; morality will still matter, but there will be no great sages or teachers. Heroic figures will still excite humanity, but ordinary life will be transformed by other forces. Beliefs that have held through centuries will be rewritten. A new society and a new age will emerge based on the foundation of equal rights. We are in the process of creating this new world. We are entering this new age. Of course, if this universal recording technology is centrally controlled, we might be confronted with the opposite of a society based on equal rights: totalitarianism.

References

Bennetts, L. "Secret oval office recordings by Roosevelt in '40 disclosed." (1982, January 14). *The New York Times.*

Haldeman, H. R. "The Decision to Record Presidential Conversations." (1988). *Prologue.*

Jones, F. A. *Thomas Alva Edison: Sixty Years of an Inventor's Life.* (2016, November). Kessinger Publishing

Montanaro, D. "The Shadowy History of Secret White House Tapes" (2017, May 13). National Public Radio.

Nixon, R. *In the Arena: A Memoir of Victory, Defeat, and Renewal.* (1990). Simon & Schuster.

Lin Ting. "Recording Archives and the American President." (2001, February). *Shanghai Archives.*

Chapter 6
Data Civilization: How Recording Technology Empowers Society, Business, and Individuals

Abstract There is a "golden thread" running through Commercial Civilization—the same as human civilization, or China. This chapter investigates statistics and recording by looking at the history of the printing press, examining the profound theories contained within A Dream of Red Mansions, and consider the development of Chinese porcelain… From history to present, from West to East, these case studies are intended to give us personal, commercial, and national strategies for the development of a great Data Civilization.

6.1 Data is the Soil to Grow a New World. If We Hope to Use It, We Must Protect It

On the same day, at various places, including the United States and China, twelve extraterrestrial spacecraft descend on the world. This causes a global panic. The aliens are sending messages but humans cannot decode them. War is on the horizon.

This is the plot of 2017 sci-fi blockbuster *Arrival*. When they realize they are not the only sentient race in the universe, humanity experiences a cosmic social phobia. Arrival envisions how we might receive a visit from an extraterrestrial civilization. This film inspired me to re-examine the most basic origins and developments of human civilization.

In the film, Louise Banks, a linguist played by Amy Adams, and physicist Ian Donnelly (Jeremy Renner) are dispatched by the American military to communicate with the aliens. The physicist quotes from the book she wrote: "Language is the first weapon drawn in a conflict." He disagrees with her conclusion, saying that the first weapon is science.

As the narrative unfolds, neither language nor science offer a solution.

Although the aliens and humanity attempt to communicate with each other, they constantly fail. There is confusion that eventually leads to panic. At the key moment when the conflict could be escalated, Louis Banks finally understands that the answer lies in the written language of the aliens. She realizes she has to study the form, color, and meaning of the symbols. She begins to decipher these ring-shaped symbols and understand the messages that the aliens hoped to deliver (Fig. 6.1).

© China Translation & Publishing House 2022
Z. Tu, *The New Civilization Upon Data*,
https://doi.org/10.1007/978-981-19-3081-2_6

In[2]:= Module[{i = 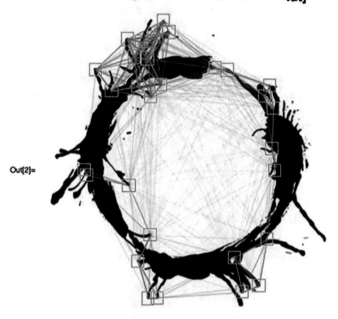 , corners}, corners = ImageCorners[i, 3, 0.1, 5];

Show[{i, Graphics[{{Orange, Thickness[0.003], Outer[If[#1 === #2, {},

{Opacity[3000 / EuclideanDistance[#1, #2]^2], Line[{#1, #2}]}

] &, corners, corners, 1]}, {EdgeForm[Green], FaceForm[],

Rectangle[# – 10, # + 10] & /@ corners}}]}]]

Out[2]=

Fig. 6.1 The written language of the aliens in Arrival (2017)[1]

Why was the solution found in the written word?

People can use their voices to express themselves, pass on information, and communicate, but the meaning can be hard to decipher. This is what Louise Banks realizes in the movie. When the aliens wanted to get their message across, they had to write it down. At that point, humans could record and decipher them. The written word is different from the voice or the language: they can not only express thoughts, pass on information, and allow communication, but they can be easily recorded. The written word allows a message to be preserved for study and analysis.

---

[1] Image from Lewin, S. "'Arrival', AI and Alien Math: Q&A with Stephen Wolfram." (2016, November 22). *Space.*

Fig. 6.2   Incan quipu, oracle bones, and Sumerian clay tablets[2]

6.2   Recording as a New Perspective on Development

Our ancient ancestors had the urge to write and paint. They did it in the sand, on tree trunks, on animal skins, and on rocks. They kept records with these methods and by another method of tying knots in rope. This is proof that our ancestors saw the need for keeping records, but it wasn't until around 3000 BC that writing appeared. In the *Book of Changes*, it says that the ancients used knots in rope and wise men of later ages kept records in writing (Fig. 6.2).

Left: When Europeans first arrived in the Americas, they discovered local Peruvian natives using the quipu system. The ropes had different colors and knots to express complex ideas.

Middle: In 1962, the Jiahu symbols were discovered at a site in Wuyang, Henan. The 17 distinct characters were written on tortoise shells and other materials more than 7000 years ago and are seen as being the source of Chinese characters.

Right: The Sumerians began writing hieroglyphs around 3500 BC. The symbols engraved on this tablet represent cows, grain, fish, and show quantities of each. This may have been the ledge of a temple. It is proof that the written word evolved from symbols.

Whether it's knotting ropes or carving a symbol, the root of all this is to record. This need to record likely came from commercial transactions. A simple system of symbols for keeping a ledger eventually evolved into a more complicated system, which evolved into the written word. The appearance of the written word shook the world. In China, legend has it that Cang Jie invented the written word. *Shuowen Jiezi*, the first comprehensive Chinese dictionary, says that Cang Jie was a historian for the Yellow Emperor and developed the written word from the footprints of birds

---

[2] Image at left is from Marshall, L. C. *The Story of Human Progress.* (1928). Middle image is from the official website of the Henan Provincial Museum. The image at right is from *History of Science and Technology.* (2004, December). Shanghai Science and Technology Education Press.

and animals.[3] Legend has it that the day the written word was invented, all the supernatural beings in the world cried out.[4]

These supernatural beings were crying out because the invention of the written word meant that humans were gaining independence. Before the written word, ghosts and spirits were spread throughout the world with the help of oral tradition. They made use of mankind's fear to rule the world. But when the written word was invented and recording began, the domain of the supernatural began to shrink, being replaced by human systems of governance. The supernatural beings that had covered the world so thickly knew that their time was limited. They were crying because they knew that they would one day be forgotten.

The appearance of the written word was the sign of civilization. As I explained before, the Chinese word for civilization is *wenming*, which combines the characters for "language" or "culture" with the character for "understanding" or "light." The "light" in this case was like the sun for the natural world—but it was lighting the human intellect. With the written word, we could break through physical boundaries, liberate our minds; we could record our observations, experiences, thoughts, and achievements in a durable, shareable form, stimulating discussion among our peers and future generations.

The word civilization refers to the time after the spread and influence of the written word; prehistory is what we call the time before the written word was invented.

But there is another viewpoint, which is that the most important starting point of civilization was not the written word but language itself. It allowed people to communicate with each other and pass on ideas. It's undeniable that this is an important step into civilization but language could only be used for communication and not recording. Language was fleeting. Experience and knowledge could only be passed on by word of mouth. Oral tradition was not a reliable way to do this. With the passing of a generation, only fragments of their experience could be passed down. Each subsequent generation would have to make many of the same discoveries already made by their ancestors. Time and energy had to be expended to replicate the efforts of previous generations, making the pace of building civilization quite slow.

Language is not unique to humans, either. As time goes by, we have more and more research that proves birds, primates, whales, bats, and even insects are capable of communication. The traditional saying is that "humans have human language and animals have animal language"; there are many legendary examples from Chinese history of people with the ability to talk to beasts. There are now examples of animal language being used in modern science and technology. One example comes from a solution to airplane bird strikes. Even a small bird becomes like a bullet when it's sucked into a jet engine. In a famous example, US Airways Flight 1549 hit a flock of Canada geese on January 15, 2009, a minute and a half after taking off from LaGuardia. The Airbus lost power in both engines and was forced to make a landing

---

[3] The original passage describing this is from the *Shuowen Jiezi*, written by Xu Zhen in the Eastern Han. The version I adapted it from is the 2015, January edition from Cathay Book Store.

[4] This is adapted from the *Huainanzi*, written in the Western Han by Liu An. I consulted a 2016, November edition from Shanghai Ancient Books Press.

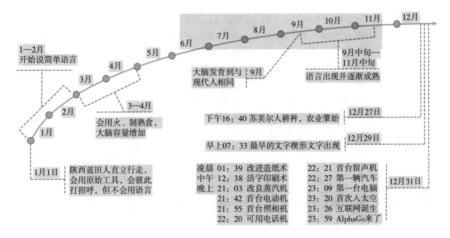

Fig. 6.3 Human civilization compressed (one day is the equivalent of 2055 years)

on the Hudson River. Keeping birds out of the path of airliners is not a task to be taken lightly. The current method involves playing a sound that will disturb birds, keeping them out of the path of aircraft.

This is why it seems clear that the fundamental difference between humans and other animals is not language. What allowed us to rise above animals was the written word. If we look at a timeline of history, it becomes clear that the written word is the most decisive invention in the rapid development of civilization. Without the written word, few of our greatest inventions could have been possible.

In order to see that development, we can condense the past 750,000 years of human civilization and chart it over the course of a year. Figure 6.3 shows that human civilization developed slowly in the absence of the written word. If human history was taking place in a single year, it would only be in the final three days of the year that the written word allowed the development of human civilization to accelerate significantly. The major events of human civilization, like the invention of papermaking, steam engines, electric motors, cars, the Internet, AlphaGo, etc. take place on what would be the last day of the year.

Truly, we can say, "without the written word, humankind was stumbling in the darkness."

The most important reason for this is that the written word makes recording possible and recording makes possible the accumulation of civilization. Civilization relies on accumulation. Every civilization becomes the crystallization of the energies and intelligences of generations. If knowledge can be recorded, we can stand on the shoulders of our ancestors. That frees people up to focus their physical and mental energies on thinking about the unknown. That is how new knowledge is created. The recording made possible by the written word accelerates civilization.

Writing not only records history and refines knowledge but also allows the rise of organization and management. When the consensus that a group has reached can be recorded, there is no more need to rely on the oral transmission of information. Instructions can accurately spread across time and space. This is when rules and laws begin to appear. These legal or administrative precedents become more durable, allowing a more advanced stage of human social organization. The ancient human civilizations that transformed their language into the written word have developed from tribes to nations, going onto a process of social division of labor and cooperation. Those groups that did not convert language to the written word remained in subsistence conditions, primitive and backwards.

Nations rise and fall; the years and then the century pass; one generation becomes the next; the living bury the dead; and all of these changes are recorded, passed on, and accumulated—and this is what creates civilization.

6.3 The Changing Fate of Chinese Civilization

China invented the written word before any other country. It is therefore one of the foundational ancient civilizations. Sophisticated characters were being carved into tortoise shells as early as the Shang Dynasty. More than 3,500 years old, these ancient marks are known as the oracle bone inscriptions.

In the Spring and Autumn Era (777–476 BC), the tortoise shells were replaced with wood and bamboo. The two words for what we might know as books are called *zhujian*, "bamboo strips" and *mudu*, "wooden tablets." These objects were still quite cumbersome and inconvenient, as you can imagine. When Hui Shi, one of the famous strategists of the Warring States (475–221 BC) wanted to carry some books along while canvassing various bureaucrats, he had to be followed everywhere he went by five cartloads of bamboo strips and wooden tablets. There's a Chinese idiom that comes from this, to the effect that someone is so learned that his "knowledge would need to be carried in five carts." Back then, if some learned man wanted to move, he would need a team of oxen to move his collection of books, which gives us another phrase: "enough books to exhaust the cattle."

In the pre-Qin period, these books made from bamboo and wood allowed China to create a vibrant culture, producing many great thinkers, and laying the foundation for a unified Chinese civilization.

In the Western Han (206 BC–AD 8), silk began to be used as the key medium for writing. That made books much lighter. But silk was expensive, putting it out of the reach of most people. Writing on silk was restricted mostly to the upper classes.

In 105 AD, the Eastern Han (25–220) court official Cai Lun built on the discoveries of earlier generations and began the first large scale production of paper.

When it came to paper, China had a long headstart over the rest of the world. In the Eastern Han, it spread east to Korea, then Japan; in the south, it entered Vietnam; it passed into western regions in the Tang, then out into the Arab world. It wasn't until the twelfth century that Europe caught up with Chinese paper production. Before paper, the medium for writing in Europe was sheepskin, which wasn't as cheap or convenient as paper.

In my opinion, the invention of paper and written language were reasons why Chinese civilization got off to a head start. Replacing bamboo strips and wooden tablets with paper allowed for the convenient recording and dissemination of knowledge. The literacy rate began rising in the Han, and there was improved social governance. China held its leading position until the fourteenth century. But, crucially, this relied on the manual copying of texts.

The symbolic event that caused China and the West to diverge was the invention and popularization of the printing press. In 1454, Johannes Gutenberg (1398–1468) gave the world movable type printing, allowing a thousandfold efficiency in copying texts and a significant decrease in the cost. Printed texts began to spread, allowing the dissemination of information and knowledge through Western societies. In the half century after Gutenberg's invention, tens of millions of books were printed, more than all the books published in the previous 1000 years.[5]

The printing press allowed the rise of science in the Renaissance. As we have already seen, science and technology relies on a medium of record to develop.

Before the invention of printing, the advancement of science and technology usually came from accidental achievements in manufacturing. Economist Yifu Lin has theorized that ancient China had an advantage in technology because of its large population, which spit out more random innovation, compared to sparsely populated Europe.[6] After the printing press, Europe began to establish the scientific method of experimentation. Accidental innovation gave way to planning. The knowledge and discoveries generated by continued experimentation easily bested the random developments generated by Chinese craftsmen.

China began to lag behind.

The international status of Chinese porcelain is an example of this. Before the eighteenth century, Chinese porcelain had a near monopoly due to advances made in production.

As Lin Yifu explains, major innovations in porcelain were accidental. The idea of glazing ceramics was discovered by kiln workers that realized coating objects in plant ash before firing would leave a glassy finish. This natural glaze made objects more attractive, made them less likely to carry contaminants, and made them waterproof. This paved the way for other attempts at manmade glazes.[7]

However, beginning in the eighteenth century, Europeans began their own experimentation, aided by comprehensive records and quantitative analysis. Their porcelain began to surpass Chinese ceramics.

Starting in 1700, the German alchemist Johann Friedrich Böttger (1682–1719) carried out more than 30,000 experiments with porcelain at a factory near Meissen. He kept complete records of each of his experiments, which means that a contemporary producer could accurately recreate his product now. Ming and Qing ceramics, on the other hand, were produced without any records, meaning they cannot be accurately reproduced. The English potter Josiah Wedgwood (1730–1795) used the same methods to carry out more than 5,000 experiments. He began with attempts to imitate

---

[5] From a Chinese translation of Febvre, L., and Martin, H. *L'Apparition du livre* (1958). Translation published by Guangxi Normal University Press 2006, December.

[6] Yifu (2007).

[7] Fukang (1987).

Chinese porcelain then went on to his own experiments. These experiments included incorporating bone ash into ceramics, producing the translucent and elegant product known as bone china.

China presents a contrast with this approach, since the porcelain industry usually involved a master orally passing down knowledge to his apprentices. The lack of records or limited records was done to protect trade secrets. The master would instruct his charges on setting up a kiln or preparing materials, but this was not recorded, which meant that apprentices had to master temperatures and amounts of materials through study and experience. There is a famous line about "ingenuity in applying tactics"—and this meant that innovation came from sudden realization, rather than being passed down from generation to generation, from day to day, or from reading records. There were certainly significant innovations among Chinese artisans but a lack of detailed records meant that they could be easily lost. Future generations could not necessarily rely on the advances of their ancestors.

My hometown of Ji'an in Jiangxi (this was known as Jizhou) was a historical center of porcelain production. Jingdezhen (also known as Raozhou) is more famous now, but Jizhou was turning out high quality porcelain a millennium ago, too. There is a famous line that "Jizhou leads and Raozhou follows."[8]

The kilns at Jizhou fell by the wayside because of a war. In 1275, the Yuan attacked the Southern Song (1127–1279) and Wen Tianxiang dispatched his troops to defend the dynasty. According to historical records, 30,000 workers were transferred from the kilns to the frontlines. They were routed by the Yuan. Many of the kiln workers that survived went to Jingdezhen.

This killed the vitality of the Jizhou ceramics industry. The fires in the kilns were extinguished. The technology developed there was spread out through Jingdezhen. This is why Jingdezhen gained a primary position after the Yuan-Ming period.

Porcelain used to symbolize and signify China, but it is now just another commodity on the international market. China still produces a lot of porcelain, but you probably won't find Chinese brands at a high-end department store anywhere in the world. Europeans hold 90% of the high-end porcelain market, with the remaining 10% mostly divided up between the United States and Japan.[9]

Another example is the study of anatomy. There is a vast gulf between thirteenth and sixteenth century European anatomy. This was not down to any major advances but the fact that printing technology after the fourteenth century allowed the dissemination of knowledge. More and more texts were printed on the subject. This increase in information would be impossible without the printing press.

Without any way to record the process, the knowledge gained from a dissection is held only by the men wielding the scalpels. Nowadays, these things are recorded not only with text but also photo, video, or even livestreams. But before the printing

---

[8] *Record of Jingdezhen Ceramic: Volume 10*, written by Lan Pu and Zheng Tinggui in the Qing dynasty contains the lines "Jiangxi ceramics were made in Hongzhou during the Tang and made in Jizhou in the Song," and "Jizhou leads and Raozhou follows." I am consulting an edition of the book published by Huangshan Publishing House in March, 2016.

[9] Jun (2014).

press, there were only limited written records and oral transmission of knowledge. This is very limiting. The next person to take a scalpel to a body would likely not benefit from the knowledge gleaned from the first autopsy.

In addition to written records, there were also systematic anatomical drawings made by physicians. If these drawings had to be traced by hand, they could never be perfect. If a drawing was copied ten times, then hundreds of copies made from those copies, distortions would be introduced.

We can see another example of this in maritime charts. It was hard for sailors to find any two maps that were the same. Since everything was copied by hand, there were distortions. Nobody was in agreement.

In the age of the print press, that was no longer a problem.

It's because knowledge could be recorded and disseminated accurately that the printing press was so significant to the development of civilization.

Many scholars have noted the power of the printing press in the dissemination of knowledge, but I think they are missing something about why this was foundational. The essence of disseminating information is preservation and recording. Manual copying of a book is essentially recording it, and the printing press speeds up that process and increases the audience. At the level of the larger society, the efficiency of recording is improved. No matter how a hand copied book is preserved, there cannot be many copies. If it is destroyed or degraded by moisture, insects, or oxidation, it disappears. A fire ripping through a library can mean a civilization rupture. But the printing press allowed a book to be printed in greater quantities and at different locations. This was a way for recording to defeat the ravages of time and allow human knowledge to be preserved and passed down on a large scale.

Compared to Gutenberg's moveable type, China's letterpress printing suffered many setbacks over its long life. Many people attempted it but they seemed to always miss the mark. During the Tang, a system of engraving wooden blocks for printing was developed. This was not unlike the idea of carving seals. A sheet of a book is carved into wood, then stamped onto a surface. Each new book has to be carved in wood all over again. This greatly increased the cost of printing. It made sense to print the key books of Confucianism but few would take the leap into the unknown involved with printing new books. A thousand years after Cai Lun's invention of paper, Song dynasty inventor Bi Sheng (970–1051) produced his own version of movable type. He was on the right track but he couldn't master the technology. One problem was that he used clay tablets, which could not be reused many times and turned out to be fragile. In the early Qing dynasty, a version of movable type printing was developed with bronze. But the technology could not keep pace. In order to print the *Complete Collection of Illustrations and Writings from the Earliest to Current Times*, 250,000 bronze movable-type characters were produced by the Qing court. After a first run of prints, the Qianlong Emperor ran into financial difficulties and the bronze printing characters were melted down for coins.[10]

---

[10] Wen (2014).

It was not until 1819, when English missionary Robert Morrison (1782–1834) printed the Bible, that moveable type printing began to spread in China.[11]

That puts China 365 years behind Europe.

Between 1454 and 1819, science and technology expanded rapidly in Europe, while China was seemingly stagnant. China in the nineteenth century was not significantly different from the Han dynasty 2000 years prior. People were reading the same books, and they were also farming the land with the same methods and tools. Pushed forward by the printing press, Europe went from a dark age to an enlightenment, building a system for storing its knowledge and history, and eventually kickstarting the Industrial Revolution. During this time, China still had not developed scientific experimentation. The key to experimentation is systematized and accurate recordings; these recordings could be passed down through generations. In China, however, knowledge was still being passed orally from teacher to disciple. This was not a problem when the knowledge being passed along were the doctrines of Confucius and Mencius; specialized knowledge, science and technology, microinnovation, and individual invention cannot be passed along in the same way. It was hard to ignite the flame of innovation. It was too easy to extinguish.

If information cannot spread and knowledge cannot be disseminated, how can society develop?

Another serious aspect of Chinese society to consider is the respect for tradition. Basically, that meant that Chinese people always considered ancient people to have been wiser than themselves. The root of this way of thinking is the lack of records.

For a long time, Chinese people believed that their ancestors knew better than them, even though their wisdom had never been recorded—or it had been lost. It is because there was no accurate record of what ancient people had done that their descendants could only guess at what they knew or didn't know. That is why the respect for tradition arose. If the ancients were so knowledgeable, what was the use of trying to invent new methods or figure out new ideas? It was better to imitate the ancients. Many reforms in Chinese history have been abandoned because they "violated the traditional order." That left the model of governance stagnant. I believe that a lack of records is at the root of this.

This respect for ancient tradition has severely restricted the creativity and progress of the Chinese people. Advances in science and technology come from experimentation and new discoveries, rather than from tradition and the words of ancient sages.

I think there's one particularly interesting example of the impact that a lack of printing technology had on China's development.

---

[11] Ruiqian (2001).

6.4   The Fate of Three Books on Three Continents

A Dream of Red Mansions is one of China's best known novels and is regarded as the pinnacle of Chinese literature.

Despite its reputation, there are many unsolved mysteries about the book. The most fundamental is about authorship: experts cannot agree whether it was penned by Cao Xueqin or Gao E. Out of all the different versions of the book, it is also unclear which should be considered definitive.

It is because of these mysteries that China has produced the field of Red-ology, dedicated to studying the novel.

According to the Red-ologists, *A Dream of Red Mansions* was likely written in the 1760s and spread through multiple private copies. There were at least 120 versions of the book, including 70 printed versions.[12] The Gengchen manuscript of 1760 is widely considered the most significant and complete.[13] It was not printed in complete form until Gao E and Cheng Weiyuan put it together in 1791 for Suzhou's Cuiwen Book House, combining the first 80 chapters with an additional 40 chapters. The 1791 print was called the Chengjia edition ("Cheng-A edition" and a second printing in 1792 was called the Chengyi edition ("Cheng-B edition").

In these 120 various versions of *Dream*, there are many errors and inconsistencies. The Chengyi edition glumly reports this fact: "The various versions of the novel create confusion. They are inconsistent and contradictory. You may find one chapter in one book and search fruitlessly for it in another. Differentiating the authentic from the false is like trying to pluck nuggets of jade out of gravel."[14] Table 6.1 contains several examples that I have drawn from Red-ologist scholars.

These comparisons between versions debate which might be more in line with the author's original vision, but they are all inconclusive. It's nearly impossible to find a version that is not contradicted by other versions. The reason for this is that Ming and Qing China lacked an effective means of recording. Books could only spread by being hand copied.

In ancient times, copying books was specialized work. The popular works that survived down through the ages only survived because someone copied them verbatim, each character, stroke by stroke… Historical figures like Sima Qian, Ban Chao, and Zhuge Liang did stints copying books. There is a poem by Master Zhu of the Song that goes: "A life spent putting ink on the page/At the end, the few hairs left are on the brush." Basically, the old men copying books had worked until they had lost all their hair and also worn out their brushes.

---

[12] Su (1981).

[13] The original Gengchen edition was compiled in 1760 and had 78 chapters altogether. Title pages contained a commentary.

[14] This is an adaptation of the original lines, which actually describe the process as one of differentiating jade from "swallow stone" (a type of stone from Swallow Mountain that was said to resemble precious stones).

Table 6.1 The similarities and differences between various versions of *Dream*[15]

位置和背景	版本	内容	评价	结论
第六回 贾宝玉初试云雨情 刘姥姥一进荣国府 背景:贾宝玉做春梦后梦遗,叫了头袭人换衣服。两人对话的内容,各个版本有较大不同。	庚辰本	袭人亦含羞笑问道:"你梦见什么故事了? 是那里流出来的那些脏东西?"宝玉道:"一言难尽"。 (宝玉告知梦中之事。) 袭人素知贾母已将自己与了宝玉的,今便如此,亦不为越礼。遂和宝玉偷试一番,幸得无人撞见。	文学家白先勇认为,袭人不可能讲"脏东西",她不了解,也没看过,心中没脏的意念。"一言难尽"不符合宝玉的口气,"偷试""幸得无人撞见"也是败笔,成了偷偷摸摸、鬼鬼祟祟的事。	此处程乙本较真实,比庚辰本更好。
	程乙本	宝玉含羞央告道:"好姐姐,千万别告诉人。"袭人含着羞悄悄地笑问道:"你为什么……"说到这里,把眼又往下里瞧了瞧,才又问说:"那是那里流出来的?"宝玉只管红着脸不言语,袭人却只瞅着他笑。 (宝玉告知梦中之事。) 袭人自知贾母曾将她给了宝玉,也无可推脱的,扭捏了半日,无奈何,只得和宝玉温存了一番。	"悄悄"两字用得好,"你为什么……"后没有话,袭人不好意思讲,对于"流出的东西",写得较含蓄。"四下里瞧""红着脸不言语""瞅着笑""扭捏"都更符合当时的情境和袭人的反应。	
第三十四回 情中情因情感妹妹 错里错以错劝哥哥 背景:宝玉挨打后,宝钗前来探病。宝玉听闻宝钗的话"如此亲切稠密,竟大有深意",又见宝钗脸红、低头、弄衣带的情态,内心暗暗产生了感慨。	庚辰本等	那一种娇羞怯怯,非可形容得出者,不觉心中大畅,将疼痛早丢在九霄云外,心中自思:"我不挨这几下,他们一个一个就有这些怜惜悲感之态露出,令人可玩可观,可怜可敬……"	"心中大畅"强调因受到宝钗关切情意的满足,不影响其形象性格逻辑的统一,符合人性和宝玉应有的情欲。"可玩可观,可怜可敬"符合对所欣赏的女性的心理情感。	此处庚辰本比程乙本更佳。
	程甲本/程乙本	那一种软怯娇羞、轻怜痛惜之情,竟难以言语形容,越觉心中感动,将疼痛早已丢在九霄云外去了。想道:"我不过挨了几下打,他们一个一个就有这些怜惜之态,令人可亲可敬……"	"心中感动"突出对宝钗情感的动心,是一种情感上的回馈,表现上明显受到封建正统观念的影响。另外,"可亲可敬"只是一种理性抽象的表达。	

Copying books by hand is quite slow, first of all. The shortest works might take a few days, but long books could take years.[16] Second, mistakes are unavoidable. That can lead to the spread of misinformation or ambiguity. To speed up the process, one person usually dictated a text while a team individually copied down their own versions. That is how many errors were introduced. It might be that the copyists misheard the dictator. Another issue was that sometimes copyists would deliberately change the original text, adding their own interpretation. Along the way, some scholars would add commentaries or insert their own notes. If someone is copying a text, it can be unclear who the original author is. These copied books don't carry metadata, such as location, date, or the name of the copyist, so there is no way for the reader to trace the authenticity of the work.

Romance of the Three Kingdoms, another classic novel, has more than a hundred versions, as well, dating back to the Ming and Qing. As with *Dream*, it can be difficult to verify authenticity among them. The reason for so many versions of *Three Kingdoms* is the same as the reason for the multiplicity of *Dreams*: the books were copied by hand. In the 1930s, the literary thinker Hu Shih summed it up like

---

[15] The table is drawn from: Bai Xianyong. Bai Xianyong Discusses A Dream of Red Mansions. (2017, February). Guangxi Normal University Press. Zheng Yun. "Comparative Approach to Various Versions of A Dream of Red Mansions." (2015, May). Doctoral dissertation, Fujian Normal University.

[16] This is suggested by examples taken from the Song and Ming dynasties in: Cao Zhi. *The Origins of Chinese Printing* (2015, April). Wuhan University Press.

this: "*Three Kingdoms* is not the work of a single author but a joint work written across five centuries."[17]

Compared with the era of writing on bamboo slips and wooden tablets, copying books on paper was an improvement. It would be impossible without Cai Lun's innovation. This was certainly an improvement, since it didn't require the direct oral transmission of information from master to disciple. In ancient times, that master-disciple relationship was often one of close personal dependency. Oral transmission relied on an intimate relationship. The teacher could decide what to pass on and what not to. With a book, the student can access information independently of their teacher—and possibly even surpass their knowledge. For a student in those times, a book could change their destiny.

The invention of paper allowed for some degree of personal liberation, but the popularization of printing magnified that effect.

Let us now turn to Europe at the same time.

Right around the time that Dream was being written, a masterpiece was published in Germany. This was *The Sorrows of Young Werther*, written in 1774 by Johann Wolfgang von Goethe (1749–1832). It caused a sensation and became a best-seller.

How was the *Sorrows of Young Werther* published? Before Gutenberg's invention, Europe was in about the same state as China in that books were copied by hand. But after 1500, publishing houses began appearing in big cities and smaller centers. Goethe began contacting specialist booksellers and went to the Leipzig Book Fair, where manuscripts were traded and sold. The Leipzig Book Fair was the largest in Germany and is still held to this day. That was where Goethe found a publisher for *Sorrows*. This created a "Werther fever" in Germany. Many young readers began emulating the novel's protagonist, wearing the same clothes, and even imitating his suicide. Goethe wrote a preface to a 1775 edition that warned readers not to imitate its protagonist. *Sorrows* swept the continent. There were six French translations between 1775 and 1778; 1779–1788 saw translations into English, Italian, and Russian.[18]

Compared to the fuzzy nature of Chinese society at the time, Europe could be seen in higher definition.

Placing these two masterpieces against each other, I personally prefer *Dream* to *Sorrows*. *Dream* is more comprehensive and profound. It's clearly the more sophisticated work. But the publishing system that produced it is nothing like the one that produced Sorrows. In Europe, a market system for books was created, allowing for a novel to spread quickly and have a large social impact. Meanwhile, *Dream* was still being copied by hand. No matter its literary achievement, its spread and influence were limited. That is only down to the social recording apparatus in the form of publishing that Europe had at the time.[19]

---

[17] Quoted in: Zhang Zhihe. "A review of research on the authorship of Romance of the Three Kings" (2002, January). *Theoretical Front In Higher Education.*

[18] Goethe (2004).

[19] Of course, we cannot discount the political environment that affected the publication of *A Dream of Red Mansions*. There are many political allusions and metaphors in the book that made it inconvenient to formally publish. But the fundamental reason was because of printing technology.

Finally, we will shoot over to North America.

In the 1770s, the continent was still a colonial territory. The best-selling book was Common Sense by Thomas Paine (1736–1809), published in 1776. Paine had come to the British American colonies from England in 1774. The book was an immediate success. 120,000 copies were printed in the first run and an additional 500,000 copies were printed over the following two years.[20] At the time, there were only two million people in North America, which meant that virtually every adult male had read the book. It stimulated the local interest in independence, and it added to an ongoing discussion about freedom, democracy, and equality. It helped push the colonies to decide to declare their independence.

Historians today are unanimous on the role the book played in American independence. *Common Sense* was a bugle, calling fighters to the War of Independence. In that war, the pen was as mighty as the sword—and perhaps mightier still was the printing press. Without *Common Sense* and the printing press, the War of Independence might have been indefinitely postponed. John Adams (1735–1826), the second President of the United States said that ideas are the forerunner of revolution; he said that the key ideas that led to the American Revolution came from Common Sense.

A book gave birth to a nation.

The emergence and spread of the print press promoted social recording apparatuses in Europe and the United States. But China fell behind because of its backwards recording methods and the inefficiency of producing, preserving, and disseminating knowledge. From around the fourteenth century, the gap began to widen in the fields of science and technology, economics, and politics, eventually putting Europe far ahead.

6.5 "Computer" Versus "Recorder": The Secret Code at the Heart of the Advancement of Commercial Civilization

As we saw in the first chapter, the new economy relies on new types of commercial practices; new types of commercial practices rely on the use of data. Data use becomes a systematized pattern. What I want to discuss in this chapter is not only the state of new business practices but the development of modern Commercial Civilization and how that was achieved through continuously expanding the scope and standardization of recording.

Luca Pacioli (1447–1517) was a pioneer of modern accounting and among the first to propose a double-entry system of book-keeping. Pacioli thought that a successful businessperson should be able to record their commercial activities to gain insight into their finances and balance their accounts. This is not as simple as recording figures on a piece of paper but involves a scientific method. The core principle of double-entry book-keeping is that a change in one account must be matched with a

---

[20] This is taken from the Wikipedia entry.

change in another account. Both debits and credits are recorded; the sum of debits made in a day's transactions must equal the sum of all credits.

This sounds simple, but we shouldn't underestimate its importance. The double-entry system of recording financial data laid the foundations for Commercial Civilization and built the prosperity of the capitalist countries. Venice and Genoa became financial centers of Europe in the century after they became the first locations to adopt this method of accounting.

Early capitalism took this system of managing companies to the level of managing states. An entire government's financial system could be tracked. A statesman finally had the ability to track the flow, increase and decrease, direction, and change of social resources. This data could be managed because it was systematized and integrated. Apart from precision management, it also promoted exchange and circulation.

Since ancient times, China has had the concept of accounting. The term for it—*kuaiji*—is mentioned in the *Records of the Grand Historian* of Sima Qian, written in the second century BC. Since the Xia dynasty of the 3rd millennium BC, there was a system by which feudal lords would pay taxes to the government. The earliest concepts of accounting came from calculating, assessing, and recording the amounts of tribute paid by the lords.[21] There is even a mountain near Shaoxing, Zhejiang called Kuaiji, so Shaoxing is usually considered the birthplace of accounting. China, however, persisted in single-entry bookkeeping until embryonic forms of double-entry accounting appeared in the late Ming and early Qing. The system wasn't introduced fully into China until it arrived from Japan in the early twentieth century.[22] At that point, the traditional methods of bookkeeping were completely replaced.

Ray Huang's ideas on "numerical management" come from looking at this historic tendency in China. His goal was to avoid traditional Chinese historiography and explain history technically. He focused on the ability to manage numerical figures. His thesis was that it was China's lack of numerical management that it began to fall behind Western civilization in the fourteenth century:

> The biggest problem is by the time Western Europe and Japan had begun organizing affairs of government in the spirit of commercial enterprises, China still had a population of hundreds of millions that were not being governed numerically.[23]

Once we entered Industrial Civilization, the impact of records became even more clear.

Typewriters began to be popularized in the United States in the 1870s. Prior to that, things like order forms, manifests, accounts, and business correspondence had to be manually written with pen and paper. The typewriter sped things up by three times.[24] That was another leap forward in recording efficiency. At around the same time, carbon copy paper began to appear, which meant that typists could make multiple copies of a text at once, greatly improving internal corporate management and external communication.

---

[21] This is taken from the *Records of the Grand Historian* of Sima Qian.

[22] Daoyang (1982).

[23] Huang (1997).

[24] Chandler (1977).

The greatest innovation of that period was the cash register. The impetus for the invention was to stop embezzlement or theft by people handling cash. In 1884, the National Cash Register Company (NCR) was founded and went on to sell 22 million of their devices over the next thirty years.[25] This changed the business environment and made NCR one of the country's most important firms. At the time, it was estimated that a sixth of major American CEOs had passed through NCR. Its status in American business could be likened to the Whampoa Military Academy in Chinese politics. It even produced IBM founder Thomas J. Watson (1874–1956).

The earliest cash registers were mechanical and did not produce usable data, but by around the First World War, they could begin recording transactions, perform basic calculations, and had rudimentary accounting functions. In 1965, IBM introduced bar code technology, pushing cash register automation to a new level.

Another important invention of the time was the punch card time clock. It recorded by punching a hole in a piece of paper the time employees entered and exited their workplace, or how long it took to complete a job. Before IBM took its current title in 1914, it was called the Computing-Tabulating-Recording Company (founded in 1911) and their main product, as their name suggests, was mechanical time recorders for employees.

Ten years ago, when I finished my schooling in the States, I went to work at an American company. I discovered that every employee had to fill out an online form called a time card management system (this is also called a timesheet system). On the form, we had to fill in the time we had spent on various work. For example, you would fill in that you spent 2.5 h on such-and-such a project, then spent 1.75 h on another project; even the length of lunch and breaks had to be entered. Based on this data, department managers could analyze the progress of projects and control costs.

Apart from keeping track of progress and costs, I once saw this data put to remarkable use in a business negotiation. The company was approached by a client to build some software but they believed that our quote was too high. When things seemed to be at a stalemate, we went back into timesheet data and showed exactly how much labor cost had gone into a similar project. We proved that our quote was completely fair and the client accepted.

This kind of system is not particularly complicated and has a history of at least a hundred years in the United States. I worked in China between 1996 and 2006 and returned in 2014, and what I saw was that very few companies had adopted the time card management system. When I asked why, some managers told me that they felt the data input by employees was unreliable. I don't believe that's actually the case. At the end of the day, they simply had no awareness of the need for mass personnel management or the importance of data.

Beginning in the 1880s, American managers had recognized the importance of records to tracking and managing business operations. In 1886, management theorist Henry Metcalfe (1847—1927) presented his work to the American Society of Mechanical Engineers and summed the situation up like this:

---

[25] Ibid.

Now, administration without records is like music without notes—by ear. Good as far as it goes, which is but a little way; it bequeaths nothing to the future. Except in the very rudest industries, carried on as if from hand to mouth, all recognize that the present must prepare for the demands of the future, and hence records, more or less elaborate, are kept.[26]

During this period, the scientific management ideas of Frederick Taylor (1856–1915) were also emerging. Taylor advocated scientific observation, recording, and analysis; labor productivity would be increased through "time and motion studies." Taylor's ideas and those of Metcalfe meshed perfectly. Metcalfe set up a system by which any actions or links between people would be recorded: "For every act or name to be recorded, there shall be a separate card; so that the cards being combined or classified, the acts or names they represent will be so too."[27]

The information from the cards could be integrated and then transcribed into a more permanent record.

Cards were once the most popular and practical recording method in the world. In the 1890s, Herman Hollerith (1860–1929) invented a tabulating machine for punched cards for use in that year's American census.[28] Every card was an individual unit, punched according to standardized rules; the tabulating machine could quickly analyze these cards and output information for a table.

These tables were themselves an important method of commercial recording. The structure of a table makes information consistent, making it more readable for managers, but also allowing for data to be extracted for recording and processing. This innovation was what allowed IBM to grow into the giant that it became. The punch tabulation machine was first used in enterprises that required analyzing a large amount of data, such as railway companies that had to deal with a mountain of passenger and freight information each day: each passenger has to be sold a ticket, have their ticket punched, and be counted; every piece of freight has a particular origination and destination, and things like transfers, tariffs, consignment fees have to be tracked. This complex and strict process requires mastering a large amount of data.

IBM got a massive opportunity in 1935, when FDR's New Deal introduced the Social Security Act and the Wage-Hour Act. These two laws mandated employers establish files for their employees and contribute to state social security based on wages and hours worked. That meant that more firms had a need for punch card machines. IBM had a near monopoly.

The punch card machine laid the foundation for the modern computer. In 1948, IBM produced its first large-scale digital calculating machine. It took up three walls in a large room and cost half a million dollars but it could perform thousands of calculations a second.

Thomas J. Watson of IBM looked at the colossus and exclaimed that the world market would probably only require five of the machines. Today, now that nearly

---

[26] Kent (1918).

[27] Ibid.

[28] For more on Hollerith, see my *The Peaks of Data.* (2014). CITIC Press. 102-110.

every household has a computer in it, this seems like an outrageous statement. But I beg to differ. It was quite reasonable.

Most people find it irrational because they ignore the difference between recording and calculating.

In the era of Watson, computers strictly performed calculations. In 1946, an early digital calculating machine was used to process American census data, and the first IBM computers were used in NASA's Apollo program. At that time, there were not many large-scale computing tasks. In fact, the same remains true.

Today, we might think of our laptops and phones as types of computers, but they do far more recording than actual computing. If you don't believe me, pick up your phone… It's ten times more powerful than a computer was in 1946, but its main functions are recording: take a picture, make a payment, post on WeChat, post on Weibo, log into a website… Rather than performing calculations in a traditional sense, these functions produce or access records.

Very few people use computers for anything but recording. Nine out of ten functions on the computer are recording functions. We should call them "recorders" instead of computers.

Commercial Civilization has progressed through the expanding scope of recording and the subsequent introduction of management models. Whether we're talking about Ray Huang's numeric management, Taylor's scientific management, or IBM's punch card machine, we can identify a common core, which is expanding recording and standardizing the information. Commercial Civilization grows out of recordkeeping. We can trace its evolution by grasping the threads of the development of recording. In fact, all of civilization is based on recording, so there's no reason that business would be an exception.

6.6 Humiliated in an Elevator

After I graduated from university in 1996, I joined the Border Defense Corps of the People's Armed Police (PAP). My provincial unit had around 20,000 members. A few years after joining, I moved from a technical post to a border management office.

In a force like the PAP, even people in technical positions have to write speeches and reports to their commanding officers. Once I moved into administration, those reports became even more important.

At the time, I was the average STEM graduate, capable of writing code but not particularly confident when it came to formal writing.

My superior was a deputy section chief with an acid tongue. Shortly after transferring, he asked when I was from. Jiangxi, I told him. He shook his head and said, "Not a lot of honest people from Jiangxi. Bad people." I wasn't good at formal writing, so he often had to edit my work. He never missed an opportunity to criticize me. "Did you actually go to university or not?" he would ask me. As far as he was concerned, he would rather deal with a local boy that had joined the force without bothering with school.

One day as I was leaving work, he followed me into the elevator, lecturing me the whole time. There were a few people in the elevator from other departments, so they had to listen to him dressing me down. He ranted about the errors in my work. "You don't understand any of this," he said. "Not a single thing." All I could do was stare at the elevator buttons and hope that it would soon be over.

Being scolded in the elevator was the push I needed. I wanted to learn to write, but I wasn't sure how. As a STEM graduate, I wasn't afraid to learn, but I didn't know where to start. I felt like the truth was buried somewhere and I didn't even have a shovel.

I found some unexpected inspiration on a business trip. The prized possession of the head of the Guangdong Border Defense Corps Sixth Detachment in Shenzhen was a thick sheaf of newspaper clippings. I was shocked. This powerful middle-aged man was in charge of 3000 men but he still had time to cut clips out of newspapers.

When I asked him, the boss smiled at the question and told him that it was a very useful exercise.

I started my own binder of clippings. I divided it into sections and started collecting various materials. I took clips from People's Daily, Liberation Army News, People's Armed Police Report, and Public Security Report. Whenever I saw a good article, I cut it out and pasted it into my notebook. After pasting it, I would take the sheet and put it into the correct category.

No matter where I went, the binder went with me. I read articles aloud to myself.

This method worked even better than I expected. My writing ability rapidly improved. Six months or so later, I got a call from my boss asking me to clarify a point. I gave my explanation and he hung up without saying a word.

It was the first time that he had not made any criticisms.

I still have those binders of newspaper clippings. I am forever grateful for having developed this habit. Many other people have the same practice. One reason that Lin Biao was such a great general was because of his habit of recording everything. I wrote about his habits in another book.[29] Sam Walton (1918–1992), the founder of Walmart jotted down notes whenever he talked to anyone:

> As soon as he meets you, he'll look you up and down, tilt his head, and then squeeze you for ideas, occasionally stopping to jot down notes, and two hours pass before you know it, and he leaves you completely hollowed out.

Walton often took out his notebook when patrolling his stores. One account has him down on his hands and knees, inspecting products stored under a table, then turning to ask an associate how he took that inventory into account when making an order. Every word the associate said was jotted down in a blue spiral notebook.[30]

The American author Jack London (1876–1916) was also a habitual notekeeper. He had pieces of paper stuffed in all of his pockets, so he could always make a note. He would constantly take out his notes to read aloud, even if he was at dinner or in bed. He famously hung clothes lines around his room so that he could peg up

---

[29] Ibid., 75.

[30] Boynton et al. (2011).

note cards. Everything he thought or said went onto note cards. These would later form the core of his work. He died without having exhausted all the material.[31] The Chinese historian Yang Kuisong has said that he keeps tens of thousands of index cards to record information.[32]

The index card is a database that we can hold in our hands.

We're talking about a way that we can record the world around us or information about other people, but there is another type of recording that is focused on our own thoughts… I'm talking about the diary. A diary or journal records and summarizes events during a day. I've kept a diary but it's been on-again, off-again. The interesting thing to me is that looking back on the past thirty years of my career, the periods where I kept a diary were the most fruitful for me, and the periods where I stopped keeping a diary tended to be slack times. The connection between progress in my career and journaling seems clear.

There are many viewpoints from history on the role of diaries. I was inspired by Qing official Zeng Guofan's letters. He wrote a diary every day, as well as reading ten pages of history and jotting down quotes. He wanted to continue without any interruption. He said that he would continue the effort even if he was traveling. He swore to never let a day go by without writing his journals.[33]

Zeng Guofan divided his diary into three categories. The first was his day's work and itinerary, noting meetings and details; the second category was notes on reading and important issues; the third category was the most casual, recording things said by guests at dinner, or topics that had sparked some inspiration during a conversation. Everything was carefully and completely recorded.

Zeng Guofan achieved many things in his lifetime and his family line produced many important people that went on to play key roles in modern China. Zeng Guofan's individual self-cultivation and practices were crucial to the success of his family.

Just like a nation or a society, a person that is committed to recording things, does it efficiently, and has a system for it will be successful. From my own professional experience, I would go so far as to say that I haven't seen anyone not succeed that had a dedicated way to record their own life experiences. It would be unthinkable to me that someone with such a system wouldn't go on to success.

The reason I bring up these examples is that I have personally realized that recording is crucial, whether to the success of an individual or to human civilization.

Comprehensive recording and complete calculation: creating the Digital Civilization.

Civilization comes from records. The printing press provided a significant leap forward in recording capability and propelled civilizations in Europe and America. Today, a new invention is going to do the same.

I am talking about the internet.

---

[31] Tichi (2015).

[32] Kuisong (2017).

[33] This note comes from a letter addressed to his comrades in *The Collected Correspondence of Zeng Guofan*.

As we have already discussed, the Internet is sedimentary—and the fragments that accumulate and consolidate to form it are data. It's like a continental shelf, slowly being built up by innumerable pieces of data and information. But the influence of the Internet will extend beyond the commercial sector. The Internet is a recording apparatus that can be applied to the entire social sphere. When the printing press appeared in the fifteenth century, its use was mostly confined to enterprises or other large organizations. There was no way for most people to run a printing press themselves. But the Internet represents a technology that is integrated into the lives of every member of society. Compared to the era of the printing press, the fact that the Internet penetrates down to every member of society means that it is tens of millions of times more powerful.

Take Wikipedia as an example. Established in 2001, it is now the world's most important encyclopedia and the largest website for knowledge exchange. In English alone, there are 5.56 million entries. This was all accomplished through the work of volunteers.[34]

Another way of looking at Wikipedia is as a systematic recording tool.

We have already discussed the April 2013 Boston Marathon bombing, which caused nearly 200 casualties. News of this event spread quickly around the world. But there were not many reporters on the ground in the immediate aftermath, so rumors started to circulate. At 7:27 that evening, Wikipedia created a "Boston Marathon bombing" entry. The content in the entry was being updated almost once a minute, despite the fact that few reporters were on the scene. Media reports began to be sourced directly from the Wikipedia entry. In a single day, there were 7732 revisions to the entry by 1705 individuals around the world. It has now become a detailed, comprehensive source of information, which includes pictures and text. It runs to 6000 words, with 355 citations. Translation of the English-language entry served as the basis for 40 other multilingual entries.

Sites like Wikipedia, designed for knowledge exchange, are based on the social recording function of crowdsourced content and funding.

WeChat, Weibo, Taobao, DiDi, and Baidu each have different functions related to social networking, e-commerce, and search, but the core function of each platform is recording. Taobao records shopping, WeChat records social networking, and Baidu records Internet records and search history, but their models all begin by storing data.

As we have already seen, at the beginning of the rise of Commercial Civilization, cards of various sorts (punch cards, time cards, index cards) were a crucial recording device. WeChat and Weibo are quite similar to these cards, even though they are digitized. It is much easier to integrate and propagate a digital card than a paper card (Fig. 6.4).

Top: A screenshot of the entry as it looked when it was created at 7:27 EST on April 15th, 2013.

Bottom: The entry as it looked in January 2018. It's divided into 10 sections, containing more than 6000 words.

---

[34] This is according to Wikipedia's own page view statistics, accessed 2018, February 2.

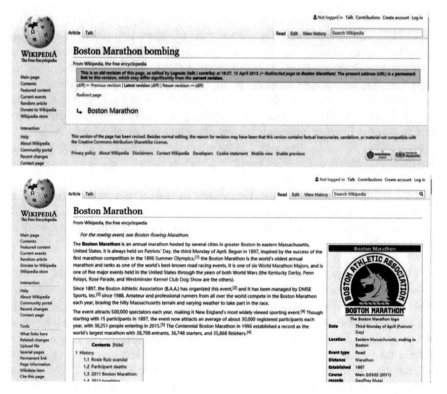

Fig. 6.4 "Boston Marathon bombing" Wikipedia entry

New online platforms like Meituan and Dianping also have a recording function. Apart from keeping track of restaurant locations, menus, contact information, etc., they also allow the public to share their opinions. This makes them a bit different from Wikipedia. Information about the world can be divided into two categories: opinions and facts. Facts would be what we would find on the "Boston Marathon bombing" entry on Wikipedia, and opinions are what we would find in Dianping or Meituan restaurant reviews. Facts are objective; opinions tend to be subjective (Fig. 6.5).

Both types of websites are becoming more important to the Internet, and that has global importance. Whether we're searching on Google or Baidu, the top results will often be sites like Wikipedia or Baidu's Baike encyclopedia, or review sites like Dianping. We've already discussed soil as a metaphor and data as the soil for a new world, and sites like Wikipedia and Dianping represent particularly rich soil. For building a new Data Civilization, these platforms will provide bumper harvests.

Looking back on the development of the Internet, the purpose of its networking seems clearly to build a systematic social recording system. What we have in the Internet is the infrastructure also for comprehensive social recording. The printing press allows the spread of information and can open up who can read certain material,

Fig. 6.5 Weibo and WeChat are the digital version of the recording card

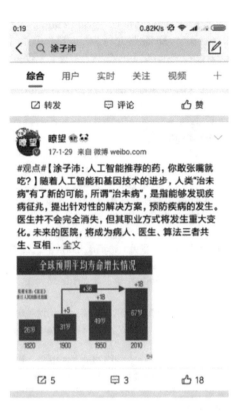

but the Internet expands that function. In the Internet age, we are not only all readers but we are all potentially authors and publishers. All of us have at our fingerprints a recording device much more powerful than the printing press.

I have stressed that knowledge comes from recording. When a society has established a recording system and the data it's recording can be analyzed, integrated, as well as freely available for access, it shines a light on the world. Fewer people will be left out in the dark. Even the most marginalized people will see benefits. Search represents a new form of calculation; a society with comprehensive recording also allows for complete calculation.

Let's say a seven- or eight-year-old comes up and asks you a question... It's very difficult to get away with trying to trick them, since they can very quickly find the truth online. That is why traditional authorities, such as parents and teachers, are being challenged today. When paper making was invented a thousand years ago, books could be written and disseminated more widely and conveniently, reducing the dependence on the oral transmission of information between teacher and student— and a similar process of liberating is taking place today. Nobody can monopolize knowledge. As long as they can use the internet to find their answer, a child can challenge their teacher's inconsistencies with the help of Google or Baidu.

The most enduring pursuit in human life is for the light. Nobody wants to be left in the dark. In the Bible, when the world is still formless and void on the first day, God commands, "Let there be light." This was a natural, physical light to illuminate the world. It became the basis for the existence and growth of all living things on Earth. The written word helped bring humanity out of chaos and barbarism. The Chinese word for civilization—*wenming*, combining the characters for "language" and "light"—suggests that it is the written word that is illuminating society.

Today, in addition to the written word, data provides another kind of illumination. In the past, different ethnic groups and regions used different scripts for their written language. These written languages illuminated their respective users' civilizations, but they did so in different ways. At its highest level, employed in powerful ways, the written word was like a torch, illuminating a nation's intellectual, spiritual, and cultural spaces; primitive, undeveloped writing is like a candle, casting limited light, which is always in danger of going out; and these torches and candles are distributed unevenly around the world. Today, data, broken down into binary 1 s and 0 s can become the universal language. The light data casts is far greater than that of the torches and candles of the written word. Data is a new sun, illuminating the entire data space.

This is a new age of Data Civilization. To adopt that Chinese word again, we could say we are going from *wenming* to *shuming*—this combines the character for "data" and "light."

Data Civilization will surpass the civilization of the written word.

Language can be used to exchange ideas and the written word can be used for recording, but humanity has already realized the limitations of a civilization built on these elements.

Language and the written word have the problem of imprecision in carrying meaning. For example, say that I tell you a certain person is sincere. The definition of sincerity is vague enough that you might have a different understanding of it than I do. Let's say I tell you a person is tall. Does that mean he's 6′3″ or 7′3″? Everyone has different standards.

Due to the different origins of various spoken and written languages, their precision can vary, sometimes in extreme ways. Cho-yun Hsu and I once discussed this problem. He said that Russian was the ideal language for poetry, while English and French were perfect for law. In his view, Chinese was somewhere in the middle, making it perfect for writing essays and prose.

Poetry requires creativity and ambiguity; the text itself is brief but deep; and legal writing requires a high degree of precision and accuracy, with clear definitions of terms that avoid ambiguity. It is said that French has the strictest lexical rules of any language, which is the reason that it is a common language of international law.

As a language, Chinese has a great deal of ambiguity. This has been examined by a great number of linguists and falls outside of the confines of this particular book. But ambiguity can produce great literary beauty; precision and accuracy are not necessarily esthetically beautiful. Beauty is inaccurate and inaccuracy is often beautiful. Nobody would describe the General Principles of the Civil Law of the

People's Republic of China as a beautiful literary work, but there is wide agreement that the *Book of Songs* is aesthetically appealing.

Also, when language and written words are used to record things, there are things that are difficult to express, scenes difficult to describe, and ideas that are not clear.

In the nineteenth and twentieth centuries, the United States had a fairly ambiguous definition of obscenity and there were frequent court cases when material was classified as obscene. In the 1964 case of Jacobellis v. Ohio, a movie theater manager named Nico Jacobellis was charged with obscenity for showing a French film called *The Lovers*. Nico Jacobellis had to take his case to the Supreme Court. They overturned the lower court's decision 6:3.

Supreme Court Justice Potter Stewart believed that the concept of obscenity could not be defined in words: "But I know it when I see it, and the motion picture involved in this case is not that."[35]

The pictures and video that Justice Potter Stewart saw are today's data. What I mean is that all products of recording, including the written word, are data. The written word is only one method of recording. If we say that the written word is specifically gold, then data is more generally metal; in this analogy, data is much more expansive and inclusive.

Data Civilization will shine brighter than a civilization based on language and the written word. This brightness will not be confined to individuals but also business and society. As we have already seen, everyone will have individual representations in digital space, which can be brought to increase the definition with which we can view society. The power of data will allow us to see society with more clarity than we could for thousands of years. Of course, there are issues with this. Data and privacy go hand-in-hand; privacy will become the main social issue. It will take a long time for the data rights to be fully established. It remains possible that autocratic or authoritarian governments could use their command of data's patterns as a tool of social command.

Looking back, it was the Internet that laid the foundation for Data Civilization and the spread of smartphones built upon that. Everyone suddenly had a recording tool in their hands that was many magnitudes more powerful than the printing press. Everyone could record whenever and wherever they wanted; we became readers and writers at the same time. Comprehensive records and complete calculations will empower everyone to build civilization. Since the Internet provides the infrastructure for the social recording, it will provide the medium for releasing the most amazing innovation since the development of the written word, reshaping human nature and the world, and building a completely new Data Civilization.

---

[35] Jacobellis v. Ohio, 378 U.S. 184 (1964).

References

Boynton, A., Fischer, B., et al. *The Idea Hunter: How to Find the Best Ideas and Make them Happen.* (2011). Jossey-Bass.

Chandler, A. *The Visible Hand: The Managerial Revolution in American Business.* (1977). Belknap Press of Harvard University Press.

Guo Daoyang. A Sketch of Chinese Accounting History. (1982). China Finance and Economy Press.

Fukang. (1987). "The Origins of China's Traditional High Temperature Glaze." In *Ancient Chinese Ceramics Research* (Vol. 4). Chinese Academy of Social Sciences Press.

Goethe. Stanley Appelbaum, translator. *The Sorrows of Young Werther.* (2004). Dover Publications, Inc.

Huang, R. *Conversations about Chinese History on the Banks of the Hudson River.* (1997). Sanlian.

Wu Jun. *Light of Culture.* (2014). People's Post and Telecommunications Press.

Kent, W. *Bookkeeping and Cost Accounting for Factories.* (1918). Sagwan Press.

Yang Kuisong. *Interviews on Contemporary Chinese History.* (2017, November). Jiuzhou Press.

Feng Ruiqian. *An Outline of Printing.* (2001, December). Cultural Development Press.

Yi Su, ed. *The Bibliography of A Dream of Red Mansions.* (1981, July). Zhonghua Press.

Tichi, C. *Jack London: A Writer's Fight for a Better America.* (2015, September). University of North Carolina Press.

Luo Wen. "The reasons why ancient Chinese moveable type was not more widely used." (2014, February 27). Netease.

Lin Yifu. "Needham's Mystery, Weber's Question, and China's Miracle: Long-term Economic Development Since the Song Dynasty." (2007). *Journal of Peking University.*

Chapter 7
New Data Governance: Building a Modern National Administrative System

Abstract How can we use big data to govern a state? That is a question that I addressed in my first two books and this chapter will elaborate on what I have learned from my observations. In the past, cities had a nervous system but no brain, but now we are entering a new age. Urban data is being integrated for AI to analyze and process. The government of the future will master data and use an intelligent network to deal with it. Compared to the traditional Chinese idea of a ruler having a mandate through virtue, the key to modernization of governance lies in "data governance". That is the only road to modernizing government.

7.1 The Road to a Better Society is Always Under Construction

In this chapter, we will cast our gaze back on the modern world, contemporary society, and China.

In November of 2013, the Communist Party's Third Plenary Session of the 18th Central Committee proposed: "The overall goal of deepening the reform comprehensively is to improve and develop socialism with Chinese characteristics, and to promote the modernization of the national governance system and capacity."[1]

As I see it, a new model of governance has to rely on data. The modernization of data collection, administration, and capacity is the modernization of state administration and state capacity.

---

[1] Decision of the Central Committee of the Communist Party of China on Some Major Issues Concerning Comprehensively Deepening the Reform. (2014, January 16 [English translation; Chinese version published 2013, November 15]). China.org.cn.

© China Translation & Publishing House 2022
Z. Tu, *The New Civilization Upon Data*,
https://doi.org/10.1007/978-981-19-3081-2_7

7.2 Digital Foundation: World Class Creativity and Local Problems

We are building a data space beyond physical space. Just as the physical space has bedrock, the data space does, too, with the digital foundation anchoring it like the pilings of an apartment block.

The key element of the digital foundation is data about individuals and juridical persons.

With a population of 1.39 billion, China is the most populous country on the planet.[2] Society is composed of people; people are the most elementary, dynamic, critical element of society. The main goal of social governance is administering and serving people. Human social activity is also the largest source of data. In *The Peaks of Data*, I described why the problem of sourcing demographic data has presented such an issue for historical and contemporary states around the world. It remains true that most countries do not have the ability to give an accurate population number. The leaders of most Chinese cities are hazy on actual demographics, especially when it comes to what is known as the floating population, which includes migrant workers.

Most urban policies are related to demography; not having accurate population numbers means that decisions are being made without a high definition picture of the city.

As a result, the idea of "close enough" rules the day.

On January 1, 2016, China began to implement a two-child policy. Many experts predicted that there would be a population surge that would bring about social problems. By 2018, it was clear that there was not going to be a population surge. In 2016, there were 17.86 million births registered, a 1.31 million increase; in 2017, that number decreased to 17.23 million, a decrease of 630,000. Expert predictions didn't line up with reality. The reason is that they did not understand demographics, like the population of women of childbearing age, and had no access to accurate numbers.

The problem is not a lack of statistics but a lack of comprehensive statistics. First, China is still undergoing a period of rapid urbanization; population flows between urban and rural areas introduces uncertainty into statistics. Second, there are multiple government departments and agencies involved in population management; departments for family planning, public security, civil administration all collect their own statistics. Having individualized data suits their own needs, but there is also no standardized cooperation or sharing. The final result is that one hand does not know what the other is doing. There is no attempt to bring statistics together. A city mayor, for example, can't get comprehensive numbers that agree with each other. They have to decide between an average value or making policies that can apply across a range of numbers.

---

[2] This is the number as of the end of 2017, not including Hong Kong, Macao, and Taiwan, from the National Bureau of Statistics' People's Republic of China 2017 National Economic and Social Development Bulletin. (2018, February 28).

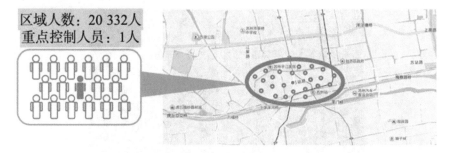

Fig. 7.1 Real-time dynamic enclosure for the floating population

The statistical problem is especially true for the floating population. If a person suddenly appears in a city, how can they be taken into account? This is a longstanding problem for city managers. Zhou Kehua, the fugitive murderer we met in previous chapters exploited these loopholes. He moved from city to city, continually committing offenses, only to be caught because of an error. Obviously, these cities could not issue prompt warnings. Zhou Kehua was a hardened criminal and was attempting to avoid detection, but what if we were talking about an ordinary person? Could they be found any easier? I don't think any municipality or government department would be capable of it, given the current level of social management (Fig. 7.1).

In the future, management of the floating population will be as easy as drawing a circle on the map. Mobile and network signals will immediately give the population number to urban management.

One day, mobile technology will make it possible.

In November, 2016, Nanjing's Xuanwu District created the first QR code house numbers. Don't underestimate that tiny square of blocks. It's perfect for a new age of data and networking. Anyone can scan the house number with their mobile phone.

Ordinary people can get basic information about the house, as well as contact information for local police, which allows for leaving messages, interactions, and online application for residence permits.

If you are the owner of the property, scanning the QR allows notification of authorities of renters or tenants.

If a municipality employee or local police officer scans the code, they can interact with a platform with personal registration, fire code, disaster prevention, and urban management information (Figs. 7.2, 7.3, and 7.4).

In May of 2017, a social governance information platform was officially launched in Nanjing's Xuanwu district.

An important idea in population management is "define the person by their dwelling," so that people, registration, and dwelling can be clearly defined and correspond one-to-one. Suzhou, Jiangsu's attempts to do this have been quite outstanding and effective. On a mobile phone screen, people, registration, and dwelling all correspond (Figs. 7.5 and 7.6).

Fig. 7.2 Nanjing's prototype QR code house number

Fig. 7.3 Scanning the QR code takes ordinary citizens to a community public services page

Staff with mobile terminals can quickly call up and verify buildings in a residential compound, as well as population. Demographic data is also displayed, divided between the floating population, the permanent resident population, and particular populations that local administrators want to keep an eye on.

Management through mobile devices improves the accuracy of population data, whether seen in short- or long-term. Rather than filling out tables in an office, workers are free to travel around the community. If you have a mobile phone, you can scan the QR code on a house number and get access to a population database. Various departments can access the same database and work collaboratively on social governance. For example, the community management of Xuanwu and the public security

Fig. 7.4 When an authorized worker scans the QR code, they can access a social governance information platform

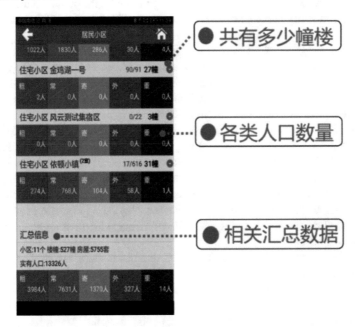

Fig. 7.5 Suzhou's population management: from the residential compound to the apartment block

Fig. 7.6 Population management in Suzhou: page layout moving from building to registered household level

management have set up a clear division on responsibility for maintaining the population database. Both sides can serve as checks and balances on each other. If they find inconsistencies, social workers can send out notices to the public security department, who can verify and correct it. Data can be modified and supplemented in real time.

QR codes are widespread in commercial applications, but this is the first use of them in population management. I believe it to be of great significance. It upgrades a house number from a multi-digit number to a public service gateway that allows for collaborative mobile data management. This is a new strategy to solve a millennia-old problem in managing population mobility. Given enough time, the quality of China's population data can be massively improved and perhaps even perfected (Fig. 7.7).

Nanjing's QR code system has been gradually adopted in other localities. In January of 2017, provincial authorities in Fujian ordered all jurisdictions to establish common standards to promote QR code house numbers.[3] Guangzhou, Dongguan, Jinan, and other centers have also begun using the QR code house number system.

Having an accurate basic population database lays the foundation for comprehensive social management. Using QR codes to manage the floating population is a world-class innovation.

When it comes to basic data collection, China has made many similar advancements, but there are still some outstanding issues to solve, such as data authenticity. Like population numbers, corporate data is another basic national requirement. However, if we look at data on corporations, it becomes clear that most municipalities have no idea even how many enterprises operate within their jurisdiction. The reasons for this discrepancy are the same as those with population data: multiple

---

[3] Regarding the implementation of standardized QR code address system. (2017, January). Fujian Province Public Security Administration.

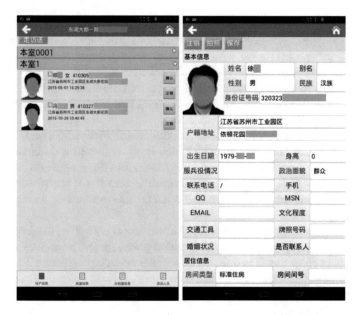

Fig. 7.7 Population management in Suzhou: page layout moving from household to individual level

departments maintain their own separate records, data is not uniform, and there is a high rate of omission in data collection. The biggest problem, though, is the immense scale of publicly available data that is completely fake.

Differentiating real data from false is the greatest challenge China faces in the field of local data collection. Building a data set with inauthentic information is like attempting construction with rotten lumber—eventually the frame starts to buckle, the bridge begins to crack.

We have one example in social security, which, according to current policy, companies must pay based on employee wages, and which is deducted by a government department based on wage information. Individual employees are responsible for about 10% of their salaries, while companies are responsible for 28%. I will use Zhejiang as an example: looking at labor market conditions, it is unlikely that many people are earning wages at the monthly minimum wage of 2819 yuan, but, according to what companies are reporting, as many as 90% of workers are making just that. The reason is simple. The higher the wages reported, the more the employee and the company have to pay. It's better to report as little as possible. Although it's also funded by individual and corporate contributions, companies will often report higher incomes to the department in charge of the housing allowance, since that money is not taxable and can be readily withdrawn and spent.

This is what leads to two government departments having data sets that don't agree with each other.

Of course, if these data sets were compared to each other, discrepancies would immediately appear. The department in charge of social security is not ignorant of this, even if they pretend to be. When they receive data from other departments, they take pains to avoid any analysis that would show these discrepancies. Crunching the numbers would be an immense task and, given the unreliability of data across the board, potentially impossible. So, they play the ostrich, burying their head in the sand, pretending not to notice the problems.

Compared to other countries, social security contributions might be a few points higher in China, but the mechanism by which it gets deducted is different. Foreign countries deduct social security based on dynamic, real-time wage information, but that adds in a new risk: if there was dynamic determination of contributions, many small- and medium-sized enterprises might not be able to handle it, leading them to close up shop.

The basic figures for the social security departments, as well as the proportion of contributions is the responsibility of the central government. The problems we have discussed here show that it is difficult to get authentic corporate data. Fake data is worse than having no data at all. Fake data cannot give us correct solutions and it can add to the problems faced. In the long run, trying to make policy with fake data is not worth the effort.

7.3 Networking Data: Holistic Governance in the Data Sphere

Issues with population management and fake corporate financial data shows us the importance of authentic, integrated information.

The 1990s was when the Internet took off, and the 2000s was when we saw the Internet of Things (IoT) begin to emerge from it. The idea was that not only would people be connected, but so would devices. At a certain point, when everything is connected, we get the so-called Internet of Everything (IoE).

As I see it, in the future, whether we are talking about people connecting to each other online, or devices linked through IoT or IoE, the final form will be the Internet of Data (IoD).

As I said in previous chapters, every person and object becomes a single point of information in the data space, with organized, structured, textured data behind it. Borrowing from the logic and concepts of IoE, this data point becomes a networked entity, or, a "data presence."

In other words, every person and object can be defined in data. Data is not everything, but everything can be converted into data. A networked node is a data presence; a data presence will have a large amount of primary data to define it but also integrated metadata; and each data presence will have its own unique data pattern. Through networking, an individual data presence can begin to be linked to other presences.

Digression: As in the old saying that carriages should have wheels of the same size, data should be standardized.

Metadata is "data about data"; it's the "foundation of the foundation" of big data. This is not the content of the data but how the data is introduced and explained. This can only be realized through standardization. A good example of this is the explanatory information on food and drug packaging. That explanatory information represents metadata.

According to this standard, much of what we call data could more accurately be termed metadata. Take a phone call as an example: the content of the conversation on the call is data, but the time the call was placed, its duration, the physical location of the two callers, etc. are metadata. Metadata is a bridge that connects data. Therefore, it must have a unified standard. That means that for any data resource, there should be only one item of metadata to correspond to it.

When metadata is standardized, data can be integrated. This is why I have used the old saying about carriage wheels and tracks across a nation being of a uniform size. The standardization of data will have the same historical significance as that measure or the idea of standardized written language. Standardization of data is not merely a slogan but a movement.

To be linked is to have a relationship. The discovery of links between pieces of data on a network will one day be automated. With a large amount of metadata, a data presence will be able to automatically discover its possible relationships to other presences on a network. These relationships, once formed, will also be automatically defined and stored. When a piece of data in a data presence is updated, the change will be pushed to other presences on the network, updating synchronously. In other words, there will be mutual, interactive updating of data, like a butterfly effect.

With networked data, the relationships between digital presences can not only be automatically established and maintained, but, when the metadata is updated, they can be automatically defined, updated, or deleted.

In other words, future data networking will create a dynamic, intelligent web that stores a large amount of intersectional information. When an individual data presence changes, other presences will check whether or not the changes are "trustworthy": if it's deemed reliable, other presences will update their data; if it's deemed unreliable, there will be no change to other data presences and the change to the original presence will also be rejected. The final result is that changes will be reconciled or rejected simultaneously, but many numbers along the chain of data will remain unanimous in their decision. This is the principle of the "distributed ledger" that forms the basis of the blockchain.

The purpose of this kind of conformity in network data is not simply to form relationships but is intended to allow simultaneous interaction and sharing between data presences.

The private and public sectors present their own particular challenges for networking data. In the private sector, there is decentralization and it's difficult to balance a wide range of competing interests. Factoring in disputes over privacy and data ownership, it becomes increasingly difficult to network data across firms. In the public sector, the main actor is the central government, which can give orders to subnational governments. There are still barriers within government, but they can be overcome. The heads of governments at the provincial and city level can send a message to jurisdictions below them, emphasizing that data belongs to the highest level of government, rather than particular departments, and it must be completely and unconditionally integrated.

The conditions are there within the public sector for the networking of data. I believe that they will take an early lead over private sector competitors.

The networking of data is a method; the final form is holistic governance of the data dimension.

Let me give you an example from policing: investigators often need to know if someone involved in a case is married and whether or not they have any property registered under their name, but this information is held by bureaus in charge of civil affairs and real estate, respectively, rather than public security organs. This information is always changing. If the person being investigated registered their marriage a day before and bought an apartment a week prior, the information has not trickled down to police databases. The current process has the Public Security Bureau putting in a data request to the relevant bureaus, which will locate and verify information before sending it over. In less developed regions, this process relies on countless phone calls.

I predict that data networking will at least double the efficiency of work at multiple levels of government. Those improvements in efficiency will be seen not only at the level of internal management but also in the provision of services to the public.

After data networking within the government and building holistic governance of the data space, this kind of situation will no longer exist. Data will be synchronized and shared to the police in real time. I worked in the government and military for ten years, so I have personal experience of data sharing issues among departments. Oftentimes, work in government departments comes to a halt while waiting for data. Employees will make use of friendships to speed up data exchange.

I arrived back in Hangzhou from Silicon Valley on the eve of Spring Festival in 2015 and went to the Yuhang District government's municipal service center to start water, power, and gas service at the place I was renting. The main hall was spacious and bright; it conveniently concentrated all the services anyone would need; and residents and workers met in an open setting, with information laid out on a desk accessible to both.

The hardware was modern, reflecting progress in the country over recent years, but I realized it was lacking software to support it.

I went to the first desk to get water arranged. The staff member checked my documents and then said that I needed to make a copy for their records. I went to the copy room, lined up, paid my fee, and made the copies. After finishing at the water desk, I looked around for the power desk and found that it was right beside me. There was nobody else waiting, so I immediately shifted over. I never expected the young woman at the desk to cheerfully request a copy of my documents for their records.

Even though the two desks were right beside each other, they were isolated in their tasks. I conducted a survey and found that it's normal for municipal service centers to operate this way: different departments are concentrated in a single space but do not communicate with each other.

If data about various utilities were networked together, it would mean being able to submit documents only once; also, by concentrating services at a single point, they could have avoided the problem of a group of people lining up at one desk, while another sat empty; and, more importantly, networking data for different utilities would improve the level of social governance and guide scientific decision-making on policy.

While I was planning the City Brain project in Suzhou Industrial Park, the city was trying to integrate provision of water, power, and gas utilities. When they began carrying out tests, they came across one house that was connected to and using utilities despite having no registered occupants and local social workers finding nobody living there. That led to another local department in charge of community management to carry out further checks that led to the arrest of two wanted criminals. Another issue was found in a particular housing complex that showed a household with two registered occupants was using utilities at a far higher rate than expected. When the police investigated, they found that the place was being rented out to unregistered foreigners. After penalties were levied against the foreign residents, hundreds of other foreigners took the initiative to register with the police.[4]

The networking of utilities data is only one example, but it's proof that the central government's efforts to integrate information will see many more innovative applications. One of the most valuable applications I have seen over the past few years is in the realm of credit: using public data, the government can create a credit platform that will rival Alibaba's Sesame Credit (Fig. 7.8).

When it comes to applying big data, there are two profitable paths for the private sector: one is in precision marketing and the other is credit evaluation. We have already discussed precision marketing, where Chinese companies are keeping pace with American firms. In credit evaluation, China is still lagging far behind. In 1837, the United States established its first credit rating agencies and 92% of the adult population of the country has a credit history. In China, only 33% of the population is connected to the personal credit system of the Bank of China.[5]

---

[4] This case is drawn from the Suzhou Development Zone Public Security Exploratory Symposium, August 27, 2017.

[5] This is according to statistics in the China Banking Association's *Blue Book on China's Financial Industry Development*. As of August 2017, the Bank of China's personal credit information system included 930 million individuals, with credit information on 460 of them. As of the end of 2017, the total population of Mainland China was 1.39 billion. Based on these numbers, 33.1% of the total population has credit records.

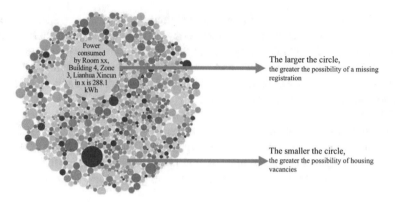

Fig. 7.8 Visualizing and predicting housing vacancies and missing personnel registration in the community via water, electricity and gas data

I have already talked about the operating mechanism of Sesame Credit, which uses a big data, but it must be pointed out that the government also has an immense trove of data, which includes not only basic individual information but also information on education, taxation, medical care, social insurance, medical insurance, and public transportation. With this data source, the government could evaluate and score citizens' credit, just like Alibaba.

Nanjing has carried out successful experiments in this field. While planning their smart city project, they integrated data from multiple departments and developed a citizen credit evaluation model, then with the cooperation of the Bank of Nanjing, the city launched the Nanjing E-Loan program, providing consumer loans to citizens, through My Nanjing.[6]

This is a benchmark in value-added government data projects. Integrating data across departments is not only to realize the goal of lifelong recordkeeping for lifelong management—but also lifelong service (Fig. 7.9).

With a sufficient level of data networking, any level of government can easily and readily create their own version of Sesame Credit to serve citizens.

Nanjing E-Loan has a lot of potential as a promotional tool. Limitations to the credit management system can restrict the development of China's commercial civilization. We are entering an era in which data is credit and credit is data. In today's China, personal data is more abundant than in any previous age. This gives us the opportunity to build a glorious and magnificent civilization.

---

[6] My Nanjing is an app made by the local government information center to concentrate various information and services. It's an outstanding example of a city government app.

Fig. 7.9 Bank of Nanjing's My Nanjing and Nanjing E-Loan

Digression: Why building a credit system might allow "overtaking on the inside".

If the foundation of human credit in Chinese tradition is human emotion and relationships, then the foundation of credit in a modern society is data. More than a century ago, when the United States began constructing its credit system, the cost of collecting data was very high, there wasn't a lot of it, and it mainly concentrated on household assets, loan payments, credit card overdrafts, utilities, lawsuits and liens, and other aspects, which introduced their own limitations. It was hard to see the "big picture" of an individual.

Big data records all aspects of an individual's personal and work life, so it's able to go beyond the "big picture" to a "3D portrait." A person's worth is determined by countless defined behaviors; individual behaviors set a person's position in the credit system. China's credit system might be able to be built even quicker and cheaper, due to the supply of data. Its multi-dimensional data evaluation capabilities will be more refined and accurate than traditional credit evaluation methods. That's why I say that the Internet, big data, and AI will allow China's credit system to overtake on the inside.

For "holistic data and holistic governance," the data is already mature and the framework is clear. Whether or not it's carried out, the fact remains that the data is there. It's an institutional rather than technical issue. Why do I say that the issue is institutional? I predict that once data is fully networked there will be clashes between existing departments over it, possibly leading to serious crises.

Dividing and handing over data is tantamount to handing over power. The networking of data will lead to the reorganization of the functions and processes of government agencies. Many government departments will be merged and others will simply be phased out. This is in keeping with the required administrative reform. Building holistic governance in the data space will necessitate a new round of reforms.

2018 marked the 40th anniversary of China's Reform and Opening. 40 years of outstanding achievements, 40 years of striding forward with confidence, facing a new tide, and now we must set out again on a new road. The road to a better society is always under construction. We need to set our eyes on the distant horizon and go there. In the physical world, our vision may be blocked by tall buildings, fog, dust, and maybe even air pollution, but in the data world, the only obstruction is data that remains outside the network.

7.4 City Brain: Our Own Era's Equivalent of Building the Nanjing Yangtze River Bridge

In 2017, I spent most of the year commuting between Nanjing, Shanghai, Suzhou, and Hangzhou. As I rushed between these cities—the most developed in the country—by road and rail, I often thought to myself, as the first generation builders of big data, what did we want to leave behind for future generations?

I often glimpsed the Nanjing Yangtze River Bridge from a distance. Before its construction was completed in 1968, the Tianjin-Pukou and Shanghai–Nanjing railway lines were cut in half by the Yangtze. When the train reached the river, passengers had to disembark and board ferries. The poet Zhu Ziqing wrote the acclaimed essay called "Retreating Figure" about his experience bidding farewell to his father at Pukou, the city across the river from Nanjing.

2018 marks the 50th anniversary of the bridge's completion—a half century soaking up wind and rain, the first great bridge designed and built by the Chinese people themselves. The bridge once served as a totem. It was the epitome and symbol of Chinese modernization. Pictures of the bridge appeared in newspapers and text books and on stamps and photo album covers, capturing the spirit of the age.

A line about the bridge goes: "A span crossing the gap from south to north/Natural barriers are turned into thoroughfares." I went to see it one night and watched it towering in the twilight. It was just as majestic as ever, but, just like Zhu Ziqing remembered his father's departing silhouette that day, it seemed to be growing old. I couldn't help but reflect on the past.

The United States has its own epoch-making bridge—Golden Gate Bridge in San Francisco. It's a symbol of the country and has appeared in many Hollywood movies. The Golden Gate Bridge is 2.7 km long, took four years to build, and cost $35 million. It was built in 1937, making it a miracle project for a country mired in the Great Depression. It inspired a generation of Americans. When I was working in Silicon Valley, I would often go out to walk its span. Set against the sky and sea, the orange steel was even more dazzling. Neither strong ocean winds nor ocean currents nor earthquakes had shaken it.

History is long; the ocean is vast. Bridges were once a symbol of mankind's dominion over nature, but in a new age, big data and AI will replace them, washing away the old urban order. Regardless of the times we are living in, people will be called upon to build bridges and pave roads—that is, to point out a clear direction forward. The question we need to answer is, what big data project could have as great a social impact as the Yangtze River Bridge?

As we discussed in a previous chapter, the Internet is now entering its autumn, and data is entering autumn, too. The application of big data has flourished. It's now carried out with a high degree of proficiency. We have precision marketing, for example, and personalized recommendation algorithms, individualized pricing, data credit, virtual currency… These are all key examples of the use of big data technology. They have opened the door to the market and spawned many great changes in a new commercial civilization.

Even still, these are universal applications meant for commercial use, available in many countries, and not particularly novel. Even as they push commercial civilization forward, they show a less beautiful side to the average person that comes in contact with them, and they create their own social problems. These big data projects can't be compared to the Yangtze River Bridge, which was planned at a national level for the well-being and safety of the masses.

If we are really looking for a New Yangtze River Bridge in the realm of big data, I believe it must be the City Brain advocated by Wang Jian, Chairman of the Alibaba Group Technical Committee.

The idea of an urban brain is not new but Chinese experiments have endowed it with a new meaning. We have taken the idea to new heights. When cities emerged in the Industrial Age, people began to consider the idea of them being capable of thought; the urban brain idea began to appear simultaneously across many countries in the twentieth century. In the 1930s, as skyscrapers started to be built, many people started to talk about these clusters of massive structures as an urban brain.

Le Corbusier said this:

> From its offices come the commands that put the world in order. In fact, the skyscrapers are the brain of the City, the brain of the whole country. They embody the work of elaboration and command on which all activities depend. Everything is concentrated there: the tools that conquer time and space—telephones, telegraphs, radios; the bankins, trading houses, the organs of decision for the factories: finance, technology, commerce.[7]

When management software systems, along with centralized databases, and command centers, people thought that was the form that the urban brain would take. In 2012, Song Junde of Beijing University of Posts and Telecommunications pointed out that, from the perspective of technology, a smart city had to bring incorporate intelligence into every aspect, and these aspects should be systematized under one

---

[7] Quoted in Fishman (1982).

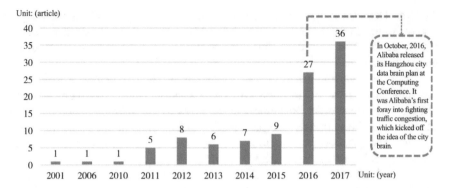

Fig. 7.10 Annual distribution of search results for the keyword "City Brain" on CNKI

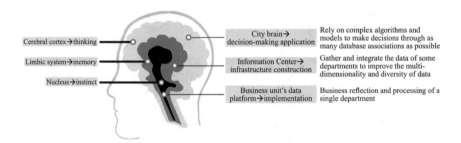

Fig. 7.11 Correspondence and analogy between city brain and Human brain

"system of systems," a central core capable of issuing commands, the brain of the city (Fig. 7.10).[8]

> The earliest discussions in Chinese of the urban brain concept on CNKI can be traced back to 2001 and a lengthy paper about IT construction in Nanhai, Guangdong. The author imagines a future "digital city" whose nervous system will be a "network of fiber optic cables" carrying all manner of information to its brain, which would be a central supercomputer.[9] Nanhai was the first county-level city in China to get online and was a pioneer in government IT. (These results only include articles and newspaper reports, not online documents.)

I think Song Junde's understanding of the urban brain concept in 2012 was fairly accurate. It is a "system of systems," after all, but technology is advancing so rapidly that this definition is no longer sufficient to cover everything the term now connotes.

I believe that the urban brain is something closer to networked data + AI (Fig. 7.11).

---

[8] Song Junde (2013).

[9] Zhang Xudong (2001).

According to researchers in bionics, we can use the structure of the human brain to inspire the construction of the urban brain; the urban brain can also stimulate the cognitive functions of the human brain. In the cerebral cortex, neurons with specific physiological function are grouped in nerve centers. There are many nerve centers in the human brain, performing their individual functions, but also working together and interacting with each other. The management of the city should also be divided and coordinated in the same way. Nerve centers correspond to departments of urban management within the government. These centers have many connections to other centers, just as urban departments must also be interconnected. Connections are made through the exchange and sharing of data. Despite this, a simple one-to-one comparison of AI and the human nervous system is too mechanical and implausible, as we have already discussed.

Let me give you an example. In August of 2017, a social worker in Shaoxing, Zhejiang discovered a group of young people had killed some eels, squeezed their blood into a water bottle, and discarded the meat. He was upset and recorded the scene. The video was uploaded to the crowd-sourced public security Qunfangyun app. A few days later, a local policeman saw the images and thought back to a few incidents he had heard of involving people faking car accidents to claim compensation. These photos of the eel got him thinking. He figured out that perhaps these people were using the eel blood to fake their injuries after claiming to be hit by cars. He investigated further and apprehended the gang.

Cracking this case relied on connecting human and machine intelligence. We could say it happened mostly by chance. Today's cities are generating a tremendous amount of data, which the human brain cannot effectively analyze and process. This is a fundamental contradiction that all cities will need to face in an era of big data.

Big data is a mountain of situational information and records. But if we collect it without the ability to analyze and judge it, we have no brain and no nervous system.

Many of our cities today are at the stage of having a nervous system but without a brain. The ability of cities to analyze and exploit data is very limited. The data dividend for cities is still budding.

Returning to the case of eel blood and fake traffic accidents, we can see that cities generate data that is unrelated. The metadata on the eel photos and the reports of the fake traffic accidents might both contain the keyword "blood," but to link these seemingly unrelated pieces of data across the Internet requires human analysis, even if they do pass through AI algorithms (Fig. 7.12).

With an urban brain, we may find other surprising links between pieces of unrelated data, like in the famous story about Walmart discovering the connection between beer and diaper sales. An urban brain can turn accidents into explicable incidents and help city managers build a society where nothing is left to chance.

The process that humanity has undertaken over the past half century began with individuals, corporations, and other organizations as their units of analysis, but we have entered a new stage, where the city is the unit. The future of urban management will use the flow of data as the foundation for analysis, discovery, and planning. I have called this "management by data."[10]

---

[10] Tu Zipei (2015a, b).

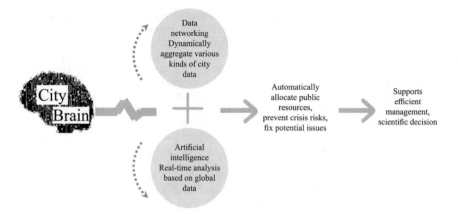

Fig. 7.12   City brain will Usher in a new era of human-machine collaboration for city management

What do I mean by the flow of data? Data is like an invisible electrical current, connecting different systems, software, and nodes. A city has the ability to record commercial transactions, the location and trajectory of vehicles and individuals. That information can then go into the cloud. This is what I mean by the flow of data. The ability of a city to use the data flow for urban governance will become an important indicator of its level of modernization. With data flow, real-time analysis and prediction will allow rapid problem solving, guide managers, avert crises, accurately allocate public resources, and optimize all decision making. We will be able to control logistics, as well as the flow of people, vehicles, and currency. This is the advanced, AI-ready form of our present day urban command centers.

This is modern governance.

The urban brain is not meant to replace human intelligence, nor is it capable. As I have already stated, artificial intelligence is not smarter than we are but simply more capable. Analyzing and grading massive amounts of data, for example, is beyond the abilities of the human brain—but not the urban brain. The analysis of AI systems can compensate for the blind spots in traditional city management. The speed at which it provides information and solutions will also be a boon. This is the advanced, AI-ready form of our present day urban command centers.

I believe that the urban brain will one day become standard in all cities in the world. It will be a new type of urban infrastructure. City management will be upgraded to a new age of human–machine collaboration. Both the 18th and 19th National Congresses of the Chinese Communist Party had proposals to "promote the modernization of the national governance system and its capabilities." From a technical point of view, the best way to modernize city management is by networking data and the urban brain system. More importantly, the urban brain concept is new, meaning that China and the United States both started from the same position. At the moment, China is already taking a lead and will have the opportunity to become a global standard bearer in the field.

Currently, only a few cities in China have proposed urban brain projects. Owing to the presence of Alibaba in Hangzhou, the municipality has made significant progress in smart city projects. But people in the industry know that the Hangzhou City Brain system is limited to administering traffic. Rather than an urban brain, it's more of a "traffic nerve center." Suzhou, Zhuhai, Hefei, Qingdao, and Macau have also proposed urban brains, but they are still in the exploratory stage. We are a long way off from realizing holistic urban governance through AI.

From the beginning of surveys in 1956 to traffic running over the span in 1968, construction on the Yangtze River Bridge lasted 12 years and required 500,000 tons of cement and a million tons of steel. The urban brain cannot be built overnight, either. It will take 5–10 years of hard work to realize the modernization of the national governance system, but those efforts are urgent and key.

7.5   One Time Only: How to Upgrade the System of "One Visit at Most"

The idea of "one visit at most" originated in Zhejiang in 2016 to refer to the simplification of the process for individuals and companies making applications to the government. The idea was that when materials were complete and there were no legal issues, that it would require only one trip to a government department.

After a year of hard work on the project in Zhejiang, it began to show remarkable results.

I live in Hangzhou and I often saw positive mentions of it on my social media feed. Figures from January of 2018 showed that 87.9% of applications were completed with a single trip, with a 94.7% satisfaction rate.[11]

The program of "one visit at most" has become a "gold-lettered sign board," announcing Zhejiang's reforms. Many other provinces are studying the program. In 2017, Shandong set "one visit at most" as their goal.

In Shandong 36,000 items for application were moved into categories of "no additional work required" or "one visit at most."[12] Nanjing went a step further, with its government service center moving to send postage paid materials to the applicant by courier.

In March of 2018, even Premier Li Keqiang mentioned "one visit at most" in a report.[13] We must pay tribute to the reforms in Zhejiang, but should they stop there? Is there room to further deepen reform and improve on "one visit at most"?

I think there is. The potential is huge.

The direction must be toward one time only—a single visit for a lifetime. I think that if we achieve holistic data and holistic governance through a foundation of

---

[11] Authoritative survey shows that 'one visit at most' measure reaches 87.9%. (2018, January 4). Zhejiang Government Service Net.

[12] Wang Chuan (2018).

[13] Li Keqiang (2018).

networking, 99% of all public services will be provided online. For most people, the situation will be not "one visit at most" but "no visits at all."

This is a matter of great significance, since reducing the number of citizens running errands at government offices will improve public services. It will also have an impact on administrative efficiency, document management, logistics, and other areas.

Firstly, the government should move the provision of public services to mobile devices. Applications should be made online. If I had to be specific, I think the basic level should be 99% of services carried out with one contact online. Currently, many parts of the country are moving toward Internet Plus government services and various departments are promoting online interaction. The problem is that they often do things according to their own rules. For example, a citizen might register one day to be connected to a public security application, but then be required a short time later to register on a separate healthcare application. Each time, the application has to be downloaded and an identity registered and verified.

Coordinating all windows on public services and their procedures according to the foundation of holistic data and holistic governance is of vital importance, which will be done through providing each citizen with a complete digital identity to carry things out through: download once, register once, and then you can access any public services or utilities.

Secondly, one of the main reasons that people have to go to government offices is to hand over documents and materials, so it's fundamental that we start reducing the amount of these. My suggestion is this: if somebody submits materials once, they don't need to do it again!

When data is networked, materials submitted by a citizen can be used by the necessary department and then accessed by other departments when they need it. The first department's audit of a document can serve as authentication for a limited amount of time, which will allow other departments to use it with confidence. In this way, citizens will submit fewer documents and materials to the government over their lifetime and be able to take care of certain applications without submitting any documents.

The meaning of "one time only" for citizens is that the government is unified: when dealing with different departments, they will not be asked for the same information. In other words, if a citizen submits their address or the names of family members to the one government department, for example, then other departments do not need to request the same data.

With this program, no government department or agency will need to repeatedly request information that is already in government databases. Citizens will only need to provide standard information once, which will be shared internally, freeing them of the burden of submitting it multiple times.

Implementing an "only once" program will mobilize the internal force of government and the external force of society to promote holistic digital governance. The government is capable of ordering that multiple departments not store their own copies of the same data, but keep it on a networked database that any department can access. This means that departments will be forced to communicate between each other, improving the efficiency of provision of services. This is a mechanism to force change.

Using this forced mechanism has proven to be the most effective way to promote reform.

Digital governance must be holistic governance.

This requires that each department must consider the government as a whole when planning IT projects. Now, many local governments have realized a system of unified planning, establishment, and review of projects, but these processes take place at the level of the project itself, proving their necessity and feasibility. I believe the process should be expanded as soon as possible to include data. Review processes should include what data is collected, so as to avoid duplication in multiple departments.

Repeated collection of the same data wastes administrative resources but because duplication can lead to conflicts in processing. If these issues remain, it will make networking data a very difficult task.

The duplication of data also increases the risk of it being leaked. We can see the danger of this in the Xu Yuyu case. In 2016, the young woman died of cardiac arrest, her health negatively impacted by the experience of being defrauded out of money she had saved for her college tuition. In applications for scholarships, she had submitted 26 pieces of personal information that were then leaked into the hands of fraudsters. She died the summer she received an acceptance letter for school (Fig. 7.13).

Left: Under the current system, the same items are repeatedly submitted to different departments.

Right: When data is networked and we achieve holistic data and holistic governance, citizens need only submit materials one time, after which it can be shared between departments.

Of course, there is a danger with the sharing of data between departments: this new level of transparency within the government might lead to staff leaking data, with data sharing making it harder to pin down the culprit. To solve this problem, blockchain technology should be used to encrypt data and record any access, modifications, or queries. With data stored on the blockchain, all of these processes can be traced and

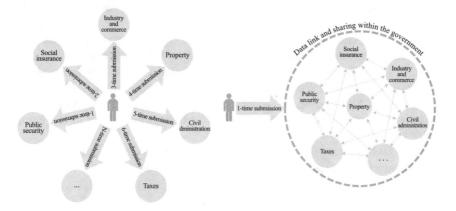

Fig. 7.13 Avoid citizens repeatedly submitting the same materials through internal sharing in the government

audited. This will be an important step in introducing the blockchain to the public sector.

7.6   Open Surveillance: The Future of Skynet

In a previous chapter, we talked about the importance of Skynet to social governance and the rapid expansion of video surveillance around the world. These cameras transform the physical world into structured image data, which accelerates the development of AI.

With fog computing, the surveillance camera is not capable of merely recording events but can also process and analyze the data it collects. The size and capabilities of these cameras turn them into something like an intelligent speck of dust. Over the next decade, these smart cameras will become the front line of AI technology developments. In short, cameras will get better and they will be more plentiful. Surveillance cameras will become an increasingly indispensable piece of urban infrastructure.

The November of 2017 child abuse incidents across China give us some idea of the importance of surveillance. Incidents at a daycare run by travel agency Ctrip in Shanghai and an operation called Red Yellow Blue in Beijing involved allegations of teachers beating children, injecting them with hypodermic needles, and force feeding them drugs. These allegations stirred up public opinion. In the incidents in Shanghai and Beijing, there was surveillance footage of abuse. At the daycare operation in Beijing, however, the footage was corrupted by a hard disk failure.

These two sensational incidents caused a chain reaction. Parents at a kindergarten in Shanxi requested surveillance footage of their own kids, but the school refused, saying that they needed to file a formal application. In the opinion of educators, the request for the video was tantamount to calling into question their reputation. Teachers vowed to resign if requests for open access were granted.[14]

The Shanxi kindergarten was not a rare case. In fact, there were similar controversies and contentions at schools across the country.

The fundamental reason is that we have no rules or clear precedents in China to follow on the opening up of surveillance. But we already have a massive surveillance network and much of it was established with public funds. Apart from the authorities installing cameras, like police forces, schools, or subway operators, who has the right to view them? Should everyone be able to view them? This is a question that the creators of the surveillance network should have been prepared from the beginning to answer.

In the United States, the UK, Canada, and other countries, the law clearly states that people whose data is held by the state or its agencies have the right to access it.[15]

---

[14] Guo Jinhui. (2017)

[15] See: United Kingdom Data Protection Act 1998, II.7.

People have the right to know which agencies have data about them and to request or make copies.[16] This type of data obviously includes surveillance data.

Digression: How American and British citizens obtain surveillance data.

Different countries have different thresholds for obtaining surveillance data. In Manchester in the UK, the rule is that people can obtain footage or photographs if they submit identity documents and specify the location of a camera and the time and date of a recording. The University of Southern California has a rule that they can provide footage in cases where these conditions are met: the victim of a crime requests video, the legal representative of a victim requests video, an insurance company involved in a case requests video, video is requested to prove personal injury or damage to property. The school also has the right to refuse the request if it endangers the safety of witnesses to a crime, it might obstruct an investigation, or the video contains materials that the department is legally unable to release.[17]

In China, the right of individuals to access surveillance footage from public areas is still rarely covered by national and local regulations. Only a few jurisdictions in China have made rules about individuals getting access to data after approval from public security officials. In Jilin, Inner Mongolia, and Xinjiang, the rules are that individuals can access surveillance data in cases where public security officials have verified the necessity. However, there are no clear guidelines on applying for that data, how public security officials should go about confirming and providing data, or the time limit for handing over data.[18]

These simplistic, broad regulations are difficult to manage and cannot meet an increasingly complex social need.

Other jurisdictions have done better in formulating regulations promoting more open access to surveillance. In August of 2016, Guangzhou's municipal legal department issued a proposal for a revised "Rules for Public Security Video Surveillance," which included the following:

In cases of infringement on the personal, property, or other rights of an individual, or in emergency cases, citizens may view relevant materials from the public security surveillance

---

[16] This is the central right afforded by the American Privacy Act, which is representative of Western regulations.

[17] The USC request site is here: https://dps.usc.edu/services/cctv-video-request/. Accessed: 2017, December 8.

[18] See: Article 27 of "Regulations for the Administration of Public Security Video Surveillance System of Jilin Province" (implemented on January 1, 2013), Article 11 of "Regulations for the Administration of Public Security Video Surveillance System of Inner Mongolia Autonomous Region" (implemented November 1, 2014), Article 16 of "Regulations for the Administration of Public Security Video Surveillance System of Xinjiang Uyghur Autonomous Region" (implemented July 1, 2014).

system, but they do not maintain the right to copy or modify the materials. Units with access to the materials must give their enthusiastic cooperation after confirming identity documents.[19]

People have the right to view surveillance footage in extreme situations, but no right to copy them. It's unclear how this regulation will be carried out. As of the publication of this book, the regulations are still in the draft stage and have not been officially implemented.

In 2015, Nanjing began preparing "Administrative Measures for Sharing and Access of Public Information Resources," but it has not yet been implemented, either. Nanjing, however, has taken great steps to increase the openness of public surveillance data.

Nanjing's innovations rest on QR codes. In the city's Xuanwu District, a QR code is affixed in areas covered by surveillance cameras. If citizens scan the code, they can access the contact information for the agency in charge of them. In the summer of 2017, when I was taking part in Nanjing's smart city project, I heard from the police about the legend of the "Chinese Auntie."

On April 6th, 2017, Auntie Li of Nanjing discovered that her electric scooter, parked around #288 Changjiang Lu, had been stolen. She went and scanned the QR code to access contact information for the surveillance camera operator. She rode her bike to the office and watched the video. She found the footage of a man pushing away her scooter. She snapped a screenshot and uploaded it to a local police app.[20] After reporting the case, two weeks went by before the police arrested a suspect surnamed Wang (Fig. 7.14).

Left: After citizens scan the QR code, they can access contact information for the operator.

Right: Surveillance image of a man stealing Auntie Li's scooter.

It's important to know that a local police station might only have ten or so officers covering a territory of hundreds of thousands of people. There might be dozens of thefts similar to the one that happened to Auntie Li. Each case requires making transcripts of statements, collecting evidence, visiting the scene, etc. Most people don't realize how time consuming and difficult it can be to investigate a case like this. When something like a scooter is stolen, the case is rarely cracked, unless the police can locate the suspect nearby. Sometimes, police arrive to investigate, see a surveillance camera, and then must figure out who administers it. It often takes several phone calls to retrieve surveillance footage, adding to the time and labor required to investigate a case. Xuanwu District now provides a way for citizens to

---

[19] "Guangzhou intends to allow individuals to access surveillance footage: experts warn of the need to strike a balance between transparency and regulation." (2016, September 2). People's Daily Online.

[20] Launched in March of 2015, the Xuanwu Police application was developed independently by the Xuanwu District branch of the Nanjing Public Security for use on WeChat. It allowed users to report crimes and file reports, a first for a police app in China. Citizens can report stolen or lost property, including official documents, laptops, electric scooters and their batteries, and mobile phones. Fraud and other cases can also be reported. After the details of their reports are verified, citizens can use the app to notify other users of crime. See: Lin Di. (2015, November 23). "Xuanwu Police launch WeChat platform." China Police News.

Fig. 7.14 QR codes for public surveillance

collect evidence by themselves and submit evidence through WeChat. Imagine how much time and effort is saved for the police!

As I mentioned in an earlier chapter, surveillance networks have become an important resource for police investigations. These electronic eyes are appearing in more and more public places. There are detractors that say these violate privacy of individuals, but, looking at it from the perspective of public security, these cameras record data that makes criminal acts more likely to be caught and prosecuted.

We are not only talking about crimes but other unacceptable behavior. When I was visiting the urban management department of one city, I heard the perfect example of this. Like many other cities, they outsource street cleaning to sanitation companies. The process is usually divided into several steps: initial spray with water, machine sweeping to clear debris, a more detailed sweeping to catch what might be caught in gutters, then a final inspection and spot cleaning. One day, the urban management department saw from surveillance footage that a street cleaning vehicle was driving its route but its mechanical brooms were raised, so that it was not actually sweeping. The sanitation company that the city had contracted with was cutting corners and faking the job. When the urban management department shared this information with the department in charge of sanitation, the latter began making use of surveillance footage. They could use surveillance to make sure the companies they contracted with were doing their job.

This is an "externality" of data.[21] By that, I mean that the way data is used might be completely different from its original intended utility, just as we saw in the origin of video surveillance in watching a coffee machine to avoid making a wasted trip. In that case, people with access to the feed could find out not only whether or not there was coffee but also who had poured the last cup. Surveillance is now so omnipresent that it can capture many things that it was never intended to be recording.

---

[21] For more on the externalities of data: Tu Zipei. (2015a, b).

On the evening of August 23, 2010, a college student named Ma Yue died after falling onto the tracks at Gulou Dajie Station. Ma Yue's mother wanted to understand the circumstances that had led to her son's death, so she requested surveillance footage. The department in charge of managing the station could not prove it. Investigators began looking into the surveillance system and discovered that data had been deleted from the hard drive due to malfunction. They found evidence of the data being accessed but nobody manually deleting it.[22] Ma Yue's mother embarked on a lengthy legal battle. In February of 2013, a court in Dongcheng District announced to her that they had managed to retrieve part of the video, but they restricted her access to it to two and a half days. They refused her requests to copy the data and submit it to outside experts for analysis.

As urban management becomes more refined and surveillance networks create more and more data, I believe that it will have to be taken out of the hands of public security agencies and departments. We can see in the above cases that it is also inappropriate for surveillance to be handed over to departments in charge of education or transit. An alternative appears in the form of big data departments that cities are beginning to set up. These departments are intended to manage urban data resources, which includes surveillance networks. My suggestion is that Skynet and related surveillance systems should be put under the administration of these data resource management departments. They should be kept independent of other departments, so that they can better serve the government and the public.

I hope at this point in the book, you have some idea of my disposition toward Skynet-type networks: I support building surveillance capabilities in public spaces. I believe the advantages to society outweigh the disadvantages. Of course, we can't deny that people walking on the street or in public parks or studying in a classroom have some right to privacy. With regards to privacy, I believe that data must be protected. Skynet should not be Big Brother but a kindly uncle, constantly watching over us, keeping us safe from street crime and accidents. We need to make sure that surveillance networks keep the truth and facts public. Individuals should have the right to access data that concerns their own vital interests.

Therefore, surveillance networks need to be increasingly open. They preserve truth and facts. They protect the safety of people and property. They can preserve the dignity of the individual. Skynet takes data from citizens, so it must use that data in accord with their desires and needs. This is the fundamental purpose that should give government data collection: lifelong recordkeeping for lifelong management and lifelong service.

The top priority must now be to make national-level policy on issues like applications for access to surveillance data, the use of that data by citizens, and the responsibilities of departments managing processes related to open surveillance.

In addition to public surveillance cameras, Nanjing has also done its best to open up access to its traffic cameras. In 2015, the city's own app began including functions

---

[22] Meng Qingli. (2013).

Fig. 7.15 Nanjing's app shows real-time traffic conditions

to access monitoring of main roads and real-time traffic conditions.[23] Citizens can use the app to see real-time images from traffic cameras on major roads, and road maps are highlighted in red, yellow, green to show congestion (Fig. 7.15).

---

[23] My Nanjing, a must-have for any Nanjing resident. (2015, April 16). Nanjing Announcements public WeChat account.

Immediately after downloading the app, citizens can access real-time images from traffic cameras.

Left: A map showing camera positions (some of them are unavailable or have low-quality streams).

Middle: Camera looking northwest toward Gulou Square on Zhongshan North Road.

Right: Looking down on Gulou Square. Zifeng Tower is at the upper left.

These open surveillance measures are undoubtedly a welcome option for Nanjing residents. They deserve careful attention from cities in China and abroad.

Who should have access to a camera? When should they have access to it? How do we prevent or solve the issues caused by opening up surveillance? These are difficult questions to answer. I believe the best step might be preparing for answers to come later. By that, I mean, for example, when installing a new camera, its capabilities and performance boundaries should be clearly shown and explained. Information must be made clear about what the camera can record, when it records, how long data from it will be stored, whether or not that data is networked, who can view it, when they can view it, and any other relevant details. This kind of clear information will become like a camera's "birth certificate." Cameras won't be installed without their certificate.

A couplet from the poet Gu Cheng goes: "The dark nights gave me my dark eyes;/I, however, use them to look for light." It's the same as with atomic energy. The power of the atom can be used in a bomb or a power plant. The former can wipe out hundreds of thousands of lives in the blink of an eye, while the latter can produce clean energy for the benefit of mankind. Surveillance networks, too, can harm people, but they can also shine new light on the world.

7.7 Regarding Modern Administration of Data and Algorithms

As I have already suggested, the Internet is entering its second face and will transform immensely. The watch word for the next phase will be data. In the first phase, the original sin of Internet companies was collecting data without consent.

In May of 2018, the European Union introduced the strictest data regulation rules in history: General Data Protection Regulation (GDPR). The GDPR is considered so strict because it goes right to the core issue of data property rights. Although the GDPR does not explicitly use that term and does not specify data rights, it implicitly introduces the concept.

The GDPR introduces, for example, the "right to be forgotten." This is a way of giving consumers data property rights without stating it explicitly. The right to be forgotten means that consumers can choose to delete data stored by a company. They have the right to have an Internet company "forget" them. The data that consumers choose to delete could be fundamental, sensitive information. This will undoubtedly have an impact on the quality of stored data. Data sets that have been corrupted have little value. This will force Internet companies to do more to show good faith and demonstrate the value of storing data. They will need to exchange tangible benefits for the privilege of preserving data. In this way, consumers can monetize their own data.

Expanding AI relies on the accuracy and completeness of data sets. Without the right data, it is difficult to develop AI. Deprived of data, AI becomes a zombie—a body without a soul.

It seems likely that with the implementation of new regulations that individual data rights will be put into practice. At that point, the game will have new rules. The original model will no longer be sustainable. Internet companies will be forced to adapt to new boundaries and give up on tried and true methods.

That's the last thing that Internet companies want. They will analyze the issue of data rights from the perspective of efficiency. In their view, it's more efficient to allocate data rights to themselves, the companies that collect data. They believe that they should maintain ownership because they have to bear the costs of collecting, analyzing, and using data to provide services. They don't buy into the argument about privacy. They don't think fear of "data monopoly" is enough reason to take data rights away from Internet companies.

This kind of thinking is very common. As we can see from Alibaba and Tencent, this model of data ownership can be extremely successful in building large Internet firms. Some observers have suggested that strict data regulations has stopped the EU from producing companies to rival American internet giants, like Facebook, Twitter, and Amazon, or China's equivalents in Alibaba and Tencent. For them, the GDPR is further proof.

As I see it, this criticism misses the mark. After nearly 30 years of wild growth, the regulation of the Internet is now being emphasized. When I talk about the Internet entering a second phase, I am also talking about the replacement of the pursuit of efficiency with the pursuit of systematized fairness and equity. To me, this seems inevitable. It's no surprise that strict regulations appeared in Europe first, given that it was also the birthplace of the modern property rights system and the contract.

Our present achievements cannot be separated from the protection of property rights. As Mencius said, "People can only have a long-term plan after knowing their property is secure." The discourse around data rights is not unexpected; this is the direction that the Internet is evolving in.

Many people are not opposed to the idea of data rights for individuals but worry that stringent enforcement will hurt the competitive foundation of Internet firms. They believe that it will hurt both Internet companies and consumers. That underestimates the ability of the market to regenerate itself. Identifying and protecting data rights will have a short-term effect, but it will lead to long-term improvements in the industry. Protecting data rights is good for the development of the Internet. It will build a solid legal foundation for the development of data security and utilities. Consumers will no longer need to worry about security issues, which will have a positive long-term impact on public confidence. This is a vote of confidence for data technologies and a necessary step in building the data space.

Whether data rights are granted to firms or individuals, this is a way of privatizing data. The GDPR is not valuable only because it's forward thinking, but because of the compromises contained within it. Drafting these regulations required threading a difficult path, balancing fairness, and making sacrifices. Compared to a plan that would prioritize efficiency, this shows great promise and high mindedness. From the perspective of our evolution as a species, we are already swimming in a data-based world. With its great bearing on the future direction of mankind, we must take these measures seriously. Big data has united mankind into a community with a shared destiny: "honor one and you honor all; injure one and you injure all." Big data might light the way to a glorious future, but not if it sacrifices our rights. If we are slaves to rather than masters of data, big data has no future.

These regulations spring from our fundamental desire to control data. Internet companies possessing a monopoly over data violates that fundamental desire. People want to protect their right to privacy and control their own destiny, rather than being shepherded by algorithms. We all hope that science and technology is being developed for the advancement of the human race, rather than to enslave us. To understand the move to regulate the internet, we must understand our own deep seated fear of big data. In essence, it is not much different from other rights movements in history.

I predict that we will see more regulations like the GDPR in other areas of the world, including China. We are about to see the Internet undergo a global governance revolution.

Data aside, algorithms will also face new regulations. We have already seen how Internet companies use algorithms and data to achieve individualized pricing, as well as dynamic pricing on ridesharing apps. When people use DiDi, they now have to contend with weather being priced into the fare, as well as surge pricing. The algorithm bumps up the price, but how exactly does it decide? If users balk at the price increase, does the algorithm know to lower it? Will it switch off the algorithm temporarily? DiDi has been forced to explain some of its algorithmic management to the public, but they don't seem to be convinced. Consumers remain skeptical.

Life in the city is now increasingly dependent on algorithms. They have become ubiquitous. Sometimes, they operate seamlessly and for our benefit; other times, they are unfair. When algorithms work against us, they can harm the public interest, increase discimination. The phenomenon of "algorithmic corruption" has appeared in recent years, by which people game the system for their own benefit.

To return to the issue of metadata, we can say that consumers are kept in the dark about metadata attached to algorithms.

Algorithms are trade secrets and they are kept closely guarded, but information about their basic operating mechanisms and efficiency should be made public. It's no different from pharmaceuticals. Drug companies don't make public how their products are developed, but ingredients and efficacy have to be made public. If we take an algorithm as a virtual product, it should come with some paperwork. We don't take a pill without reading directions, so why should we use algorithms without knowing how they work (Fig. 7.16)?

The images above are for illustration purposes only and do not reflect real world uses.

We must consider the impact of algorithms on our lives. They can even manipulate our mood and psychology. I made the same comparison in the first chapter of this book: an algorithm is like a drug. Before a drug is approved by regulators, the

Fig. 7.16 The metadata of algorithms should be made public like drug information

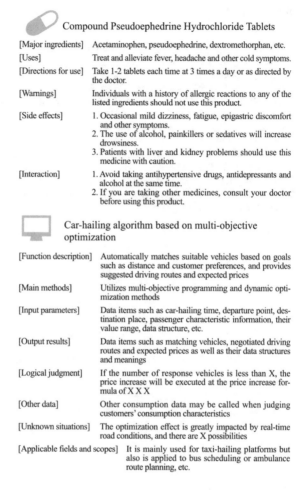

Compound Pseudoephedrine Hydrochloride Tablets

[Major ingredients]	Acetaminophen, pseudoephedrine, dextromethorphan, etc.
[Uses]	Treat and alleviate fever, headache and other cold symptoms.
[Directions for use]	Take 1-2 tablets each time at 3 times a day or as directed by the doctor.
[Warnings]	Individuals with a history of allergic reactions to any of the listed ingredients should not use this product.
[Side effects]	1. Occasional mild dizziness, fatigue, epigastric discomfort and other symptoms. 2. The use of alcohol, painkillers or sedatives will increase drowsiness. 3. Patients with liver and kidney problems should use this medicine with caution.
[Interaction]	1. Avoid taking antihypertensive drugs, antidepressants and alcohol at the same time. 2. If you are taking other medicines, consult your doctor before using this product.

Car-hailing algorithm based on multi-objective optimization

[Function description]	Automatically matches suitable vehicles based on goals such as distance and customer preferences, and provides suggested driving routes and expected prices
[Main methods]	Utilizes multi-objective programming and dynamic optimization methods
[Input parameters]	Data items such as car-hailing time, departure point, destination place, passenger characteristic information, their value range, data structure, etc.
[Output results]	Data items such as matching vehicles, negotiated driving routes and expected prices as well as their data structures and meanings
[Logical judgment]	If the number of response vehicles is less than X, the price increase will be executed at the price increase formula of X X X
[Other data]	Other consumption data may be called when judging customers' consumption characteristics
[Unknown situations]	The optimization effect is greatly impacted by real-time road conditions, and there are X possibilities
[Applicable fields and scopes]	It is mainly used for taxi-hailing platforms but also is applied to bus scheduling or ambulance route planning, etc.

manufacturer is required to define its composition, form, function, method of administration, and all possible side effects. Developers of algorithms should also have to lay out the elements, properties, and functions of their product. Internet companies should be forced to explain basic metadata, like input and output parameters. As with pharmaceuticals, any claims should be independently verified.

In traditional Internet regulation, it's content that's censored, but in the future, it will be algorithms. This kind of regulation requires personal and institutional expertise. I predict that many countries will move to incorporate this kind of expertise into government departments. In China, it could be handled by the State Administration for Market Regulation.

References

Fishman, R. (1982). *Urban Utopias in the Twentieth Century*. MIT Press.

Guo Jinhui. (2017) The anxieties of a 'high-end' parent: teachers vow to resign if surveillance footage is shared. China Business Network.

Li Keqiang. (2018). 2018 Government Work Report.

Meng Qingli. (2013) Subway company advises mother of university student that fell to his death that it was her son's fault. *Yangcheng Evening News*.

Song Junde. (2013). We need to do serious work on building the digital city, the wireless city, the smart city. Sohu Blog.

Tu Zipei. (2015a). *Big Data*, Third Edition. Guangxi Normal University Press.

Tu Zipei. (2015b) Externalities turn data into 'data blade.' The Paper.

Wang Chuan. (2018). 90% of services moved to 'one visit at most' by year-end. Dazhong Daily.

Zhang Xudong. (2001) Digital urban management: the example of Nanhai, Guangdong in IT transformation. *IT Management World*.

Chapter 8
The Boundaries of Artificial Intelligence: Risks and the Future

Abstract Data records the past, but it can also show us the future. If we have enough data, could we predict everything that is going to happen? In the age of quantum physics, this question has already been answered. In this chapter, I advocate the use of quantum thinking to examine society and propose that the uncertainty principle affects big data. Beyond that, we know that humans are incapable of recording everything, so time and space are distorted or replicated. But time and space cannot be perfectly replicated. The smart society has its charms, but it also has many limitations, weaknesses, and even great risks.

8.1 If Human Lives Can Be Predicted with Absolute Accuracy, then We Have Become Machines

In November of 2016, Jack Ma, speaking in Shanghai, suggested that over the next thirty years that data would allow the expansion of the planned economy.[1]

That was not the first time that Jack Ma had applauded the planned economy. As early as 2015, in an interview with the Korean newspaper Joongang Ilbo, he claimed that by 2030, the planned economy would be the superior system. In 2017, he repeated his claim, saying that big data and its predictive powers would redefine the meanings of planned economy and market economy.[2]

Jack Ma's persistence here has led to heated debate between business and academia. As the term suggests the past era, "planned economy" has been swept into the dustbin of history, while the term "market economy" remains popular. The return of "planned economy" to popular discourse has become a trend.[3]

---

[1] Jie (2016).

[2] Jack Ma (2017).

[3] In fact, Jack Ma is not the first person to draw links between big data and the planned economy. If you enter the term into the Baidu search bar, you will find that it was a hot topic five years before he mentioned it. In July of 2014, I was invited to speak about the issue of big data saving the planned economy at an event at Shanghai Jifeng Bookstore with "Nanqiao Xiansheng."

© China Translation & Publishing House 2022
Z. Tu, *The New Civilization Upon Data*,
https://doi.org/10.1007/978-981-19-3081-2_8

Surveying the past century, most economists believe that the planned economy was a flawed system to allocate resources. But, having said that, I believe you can already sense what Jack Ma is getting at with his statements. The Internet at present is like a supernatural eye, capable of seeing into the desires of users, so what if its big data forecasting powers were put to use by economic planners? Why shouldn't we use big data for economic planning and regulation?

Fundamentally, all human effort is about predicting the future. If you can predict the future, you can control the present. In that context, how should we view the predictive power of big data? What can big data be used to predict? What are the boundaries? If it can't be used for prediction, then what is behind its promotion? What are the risks and boundaries of an AI society?

We must have a clear understanding of these issues.

8.2   Universal Prediction and Laplace's Demon

'Tis late; the astronomer in his lonely height,
Exploring all the dark, descries from far
Orbs that like distant isles of splendor are,
And mornings whitening in the infinite.
Like winnowed grain the worlds go by in flight,
Or swarm in glistening spaces nebular;
He summons one disheveled, wandering star,—
Return ten centuries hence on such a night.
That star will come. It dare not by one hour
Cheat science, or falsify her calculation;
Men will have passed, but watchful in the tower
Man shall remain in sleepless contemplation;
And should all men have perished there in turn,
Truth in their stead would watch that star's return.

This is a poem called "The Appointment" by Nobel laureate Sully Prudhomme (1839–1907). In his bold, expressive, expansive verse, he is praising the unbelievable predictive powers of astronomers.

The ability of mankind to predict the future was the first breakthrough in the natural sciences. We know the ignorant state that our ancestors were in—unable to benefit from their ancestors and unable to pass on their knowledge. They had no idea of what was coming. Human lives passed without leaving any mark. It was only with comparatively recent advances in science that our situation began to show a glimmer of hope.

We must give credit for the establishment of modern science to Sir Isaac Newton (1642–1727). In his 1687 Philosophiæ Naturalis Principia Mathematica, Newton proposed three laws of motion and established classical mechanics. These three laws meant that the motion of objects could be explained with scientific theories. What that meant was that given the position of an object, the laws could be used to calculate and describe its trajectory by analyzing the environment and forces acting upon it.

We return to trajectories again! This time, however, we are not talking about deriving a trajectory by looking at records but predicting a trajectory through calculations.

Newton turned his gaze skyward and used the heliocentric model of Copernicus, Kepler, and Galileo to prove that heavenly bodies follow the same laws as those on Earth. The laws held on Earth and in heaven. The entire universe began to take on a different shape.

What Newton did was to express the laws of nature that govern the entire natural world. His theories became the foundation of physics and astronomy, as well as modern engineering. His scientific discoveries paved the way for the First Industrial Revolution and the entry into the age of steam and mechanized spinning and weaving.

Newton was given a state funeral. The poet Alexander Pope wrote this inscription for him: "Nature and Nature's laws lay hid in night: God said, Let Newton be! and all was light."

By the end of the nineteenth century, a century on from the death of Newton, physics was being perfected using his theories. It was discovered that all physical phenomena could be described and explained through Newtonian physics and the electromagnetic theory of James Maxwell (1831–1879). Some people even began to ponder the end of science. Perhaps, they thought, all the laws of nature have been discovered.

The methodology and theories of classical physics helped guide the course of civilizational industrialization. The core of this was objective, mechanical, precise methodology, which treated the world as like the workings of a clock. The parts of a clock are precisely mechanically connected. The trajectory of an object in space could be analyzed objectively. However we might view the world, it had no impact on the laws of physics.

This inspired psychologists, economists, and sociologists to attempt to decide that human society might also be governed by laws like those in physics. Those laws remained to be discovered. The goal of social scientists was to determine what laws might govern social phenomena.

Pierre-Simon Laplace (1749–1827) was a French follower of Newton. At the École Militaire, he had been Napoleon's teacher and would go on to serve as Minister of the Interior in his government. In 1773, Laplace applied Newton's laws of universal gravitation to the solar system, answering a popular scientific question of the time: Why is Jupiter's orbit shrinking, while Saturn's is constantly expanding? After that, he turned to social questions. In 1781, he began looking at the sex ratio of babies born in Paris. He used methods from classical physics to draw the probability of the ratio and was surprised to find that this problem seemed to be governed by the same balance and symmetry of natural phenomena. He believed social and natural problems both had universally applicable laws. In 1814, he proposed:

> We may regard the present state of the universe as the effect of its past and the cause of
> its future. An intellect which at a certain moment would know all forces that set nature in
> motion, and all positions of all items of which nature is composed, if this intellect were

also vast enough to submit these data to analysis, it would embrace in a single formula the movements of the greatest bodies of the universe and those of the tiniest atom; for such an intellect nothing would be uncertain and the future just like the past would be present before its eyes.[4]

This intellect he describes is now called Laplace's demon.

This is the peak of determinism. According to the theories of Newton and Laplace, everything, including human beings, could be governed by natural laws. In this way of thinking, human thought, behavior, and emotions run like clockwork, following a predetermined trajectory. Everything, the determinists said, was fixed and predictable. With enough data, everything could be predicted.

The reason given for why we can't predict everything is that we don't have extensive enough records and sufficient data. That matches up quite closely with what Jack Ma has claimed.

8.3   Quantum Thought: The Reality of Uncertainty

In the twentieth century, scientists set their sights on the microscopic level. They began studying molecules, atoms, and elementary particles. But they began to discover more and more phenomena that couldn't be explained by Newtonian physics. The edifice of classical physics began to crack. At that moment, a number of scientists, including Einstein, Planck, Bohr, and Schrödinger, began to fearlessly break from tradition to create the concepts and theories that would undergird the field of quantum mechanics.

In the decades that would follow, quantum mechanics would emerge as a discipline key to the study of the laws of motion and particles.

When we talk about the quantums in quantum mechanics, we are not talking about the particles themselves but states that they might exist in. For example, imagine the numerical value assigned to an object's measurement… Let's say these measurements go 1, 2, 3, 4. We can't continue arbitrarily between 1 and 2. So, 1.11, 1.111, etc. Those are quantum states. The smallest unit is called a quantum. In the natural world and social life, there are many phenomena that can be measured but when we attempt to record those results, we observe consistent and inconsistent results. For example, an accelerating car will not jump immediately from 0 km per hour to 100. There is a gradual acceleration from 0 to 100, which means that there will be a consistent state of change. Another example, this time from social science: when conducting a census, it is only possible to use a single individual as the smallest unit, so there will be no results like "1.23 people." That kind of inconsistent change is quantization.

You can understand it like this: when you get down to the smallest, indivisible elementary particles, its state cannot be determined by mankind.

---

[4] Laplace (1951).

The German physicist Max Planck (1858–1947) was the first to hypothesize that a particle's energy could not be measured continuously, like a classical wave. In addition to this, Werner Heisenberg (1901–1976) later proposed the uncertainty principle. He believed that microscopic phenomena could not be observed or measured without adding interference. In any attempt to carry out observation and measurement of a particle, the act of observing has an effect. This interference will make it hard to make key determinations. In other words, the observer themselves change the results of any observation, recording, or measurement. Therefore, quantum theory holds that microscopic particles can jump between spaces without consistent motion. Even if we have all available information and data, we cannot predict the motion and orientation of a particle. In attempting to make better measurements for our forecast, we change the behavior of the particle. Here, we have an essential disagreement with the idea of Laplace's demon.

In order to deal with uncertainty in measuring the motion of particles, quantum mechanics gives us the idea of probability. There is no way to be sure of the position or state of a particle. All we can do is use an equation to assess the probability of an electron's position, t, at a certain time, r.

The most important characteristics of classical physics are that the conclusions are objective, mechanical, precise, all summed up in the deterministic idea of Laplace's demon. Quantum theory combines the subjective world and the microscopic world. The two are inseparable. What we observe about the microscopic world must be perceived, measured, and recorded, but, by doing that, we distort things. At the microscopic level, the world is uncertain. Take for example Schrödinger's cat. Until we open the lid, the cat in the box must be considered simultaneously alive and dead. Nothing is certain.

This is quite similar to observations at a more human level. As we saw in previous chapters, a historian picking up a brush or a journalist pointing a camera will change the language and behavior of the imperial sovereign or the president. Nixon's chief of staff admitted that he spoke differently when he knew the tape was running. His ideas were being distorted by the knowledge of being recorded. People that know they are being observed and recorded will modify their behavior, even if they are not aware of it. That makes it difficult to predict their next action or plot their trajectory.

Just like elementary particles, human beings are hard to record and observe with certainty. By observing life, we change what we are observing. Economics, sociology, and philosophy have all been influenced by ideas from quantum theory.

8.4 Human Being Are the City's Uncertain Elementary Particle

We need to better understand the city, the place where most of us live. Everyone understands the city—but they don't. Sometimes, when my plane is coming in for a landing, this strange feeling comes to me with amazing clarity.

The first thing you notice as the plane approaches a city is the border between urban and rural areas. Roads cut across the land; the land is covered with farms, houses, water, and mud. As the plane gets lower, you notice highways, apartment blocks, interchanges; the land changes from dirt to concrete. At that moment, you realize clearly that the city is an artificially constructed environment. It must follow geographical features, but it's otherwise a human creation. We lay steel, glass, and concrete across the surface of the Earth, so that the city appears like a protrusion on its surface.

The plane gradually descends and speeds over the surface of the Earth. The tallest buildings look no bigger than matchboxes. The highways and the vehicles on them look like toys. The city looks like a sandbox. But you might find yourself surprised when landing to realize that it has the capacity to feed, clothe, house, entertain, infuriate, gladden, and sadden millions of people. The city is like a giant container for human life. We design it, we control it, and we try to create a safe, stable, comfortable, prosperous life within it.

The city is the greatest transformation that the human species has carried out on the planet. We live within it and also participate in its ongoing design and transformation. In the sixteenth century, before artillery made them obsolete, major cities invariably had walls. At that time, cities were even more like sealed containers of life.

Containing people within the city is the greatest transformation that mankind brought to the natural world. As I discussed in a previous chapter, human civilization comes from records, but, other than records, we must also take into account that the place where civilization concentrates is in the city. Words and data can record thoughts, but cities are where they can take form.

From agricultural to industrial society, and then on to the information society, we have continued to design, cultivate, control, and cleanse. By planning new systems and regulations and inventing equipment and facilities, we are trying to grow the ideal garden within our sealed containers.

Take individual members of society as the elementary particles in the gigantic structure of the city, constantly in motion… Our task is to study these particles. We must find a way to portray and express their behavioral choices, and also their trajectories through space and time.

In the previous chapter on the urban brain as the new Yangtze River Bridge, we saw that one new contradiction of our current system is that too much data is being generated. There is no way to harness enough manpower to deal with them. We need new plans. This data represents many different types of information, including administrative information for the government, demographic surveys, consumer and social information from the Internet, surveillance footage, etc. This is a flood of data, but no matter how it is networked, it lacks consistently and is disconnected. Like the elementary particles at the quantum level, the data jumps around. Data records one fact but must be thought of as a fragment.

I believe these fragments of data are comparable to quanta. For example, data from a surveillance camera is not showing us continuous information: a vehicle driving through the city is captured on cameras at various positions, which is converted into data, which creates a trajectory based on inference. If a vehicle appeared on Camera

A at 10:00 and Camera B at 10:30, the trajectory can be estimated, but not with certainty, since there will be multiple routes between Camera A and Camera B. We don't know if the driver might have had a small issue that caused him to deviate from his course. Maybe they took a shortcut that we couldn't predict.

Even with Skynet and City Brain, the data being analyzed lacks internal consistency and continuity. Just like electrons or neutrons, we can't make any statements with absolute definitiveness but rather statements of probability.

Imagine a pool table. The cue strikes a ball. Newtonian mechanics can calculate the trajectory. It can also calculate what will happen with the next two or three balls that are struck. But as complexity increases, calculations are increasingly difficult to make with accuracy. Now, imagine that we have a table capable of holding millions of balls… Once these balls are set in motion, this pool table is as complex as the shallow container of the urban environment. At the microscopic level, the motion of these balls is chaotic, just like people moving through the city. If we look at the system as a whole, we can start to see some patterns emerge. Only in rare circumstances are those fluctuations particularly large.

An urban management team does not need to know the trajectory of every one of these particles. If we return to the pool table analogy, they would only need to know general ideas, like when a ball went into a pocket, hit a certain point on a bumper, which balls it collided with, etc. With that information, trajectories can be plotted. For locations outside of those about which specific information exists, probabilities can be calculated.

This is how data is now collected, used, and analyzed for urban management.

The trajectory of people through the city is as difficult to measure as the motion of elementary particles. As we have already seen, cameras and tape recorders change a person's behavior. But there are other ways to change behavior through data collection and information dissemination. We have an example in the traffic advisory boards now installed on roads. They are color-coded: red indicates congestion at a certain stretch of the road or intersection, yellow is busy, and green is light traffic. You might choose to take the road highlighted in green, but you'll quickly notice that the traffic is picking up. Why would that be? Many other drivers will have noticed the traffic advisory and made a similar decision.

We can take the stock market as an example, too. At any moment, there are many people buying stocks, but imagine that an individual buyer has been granted the ability to predict the future price of a stock. If this individual knows the stock price is going to shoot up, he will buy a large number of shares. The problem is that his large purchase will not go unnoticed. He will affect the market. Individuals, the elementary particle of this particle marketplace analogy, will change their behaviors. They might decide to buy or sell in large quantities. The entire stock market will be influenced by the purchase of the first prophetic buyer, influencing the trend. But if the prophetic buyer knows the stock is expected to shoot up and does nothing, his powers are for nothing.

This is the pitfall of trying to predict human behavior. If we can predict a result, it will change the outcome. No matter how accurate, once we try to act on the prediction, we find it impossible to actually do anything.

There is an ancient fable in India about a Brahman priest attempting to disprove the Buddha's predictions. He took a sparrow, clutched it in his hand, and draped a long sleeve to hide it. He asked the Buddha whether the bird was alive or dead. If the Buddha said the bird was dead, the Brahman priest could take it out and prove him wrong. If the Buddha said the bird was alive, the Brahman priest could pinch its neck with his fingers, then take it out and show a dead bird. The Buddha declined to answer. He walked slowly to the threshold of the temple gate, paused, and asked, "Am I going in or going out?".

The Brahman priest could never give a correct answer. As soon as he made his prediction, the Buddha could do the opposite. If he kept his prediction to himself, he has a 50–50 chance of getting it right, which is no better than taking a random guess.

Human beings have free will. They are particles whose state cannot be predicted. We can never make accurate predictions about human futures because somebody will invariably come along and do the opposite. The prediction itself will change the behavior of the person whose behavior it attempts to forecast. For this reason, individual behavior cannot be predicted. When we are talking about groups, though, it becomes more feasible. Individuals have free will but they are balanced out within a group. That is why it's hard to predict the price of a single stock but much easier to predict general market trends (Fig. 8.1).

General predictions can be made, but only in situations where the target of the fore-cast is unaware of their behavior being recorded, observed, and estimated. The target must be completely undisturbed. This is almost impossible because maintaining records and gathering data is itself a type of interference.

Apart from this fundamental problem of predicting human behavior, there is also the limitation of collecting data. The knowledge granted the ideal intellect proposed by Laplace is not possible to gather about an individual. This means that it is almost impossible to make predictions about social and economic affairs with any degree of accuracy.

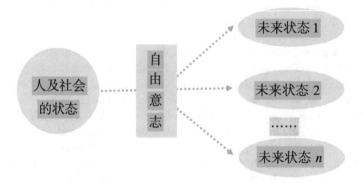

Fig. 8.1 Human beings have free will

8.5   Data Relativity: The Limits of Universal Recording

The source of data is recording.

Records have no boundaries. As Su Shi writes in *The Red Cliff*: "Only the cool wind on the river, or the full moon in the mountains, caught by the ear becomes a sound, or met by the eye changes to color; no one forbids me to make it mine, no limit is set to the use of it…" The meaning of this is that there is no limit to the things that we can record. Even if we wanted to, we would never run out of things to capture.

There is no limit to what can be recorded, but there is a limit to our capacity to record. We must admit that there are still many things that we are incapable of recording.

American novelist Nathaniel Hawthorne (1804–1864) has an interesting story that shows us the limits of records. It also has lessons about what we would face if we did break those limits.[5]

David Swan is a handsome young man on his way to Boston. While waiting for the coach, his exhaustion overcomes him and he falls asleep under a tree.

After sleeping for a while, a wealthy husband and wife encounter him. Their carriage has broken down. They have recently lost a son and have nobody to inherit their fortune. When they see David, they are taken with him. He has an infectious allure. They note that he is young and healthy. They want to get to know him. They decide they could pass their wealth on to him.

They say that it's as if God laid him in their path. But just as they're about to wake him up, someone rushes up to tell them that the carriage has been repaired and they can get back on the road. They are shaken from their reverie and realize the absurdity of passing on their inheritance to a stranger.

After their carriage leaves, a young woman arrives. When she sees the sleeping young man, she has the feeling of having stumbled into his bedroom. She blushes with embarrassment. She lingers for a while and shoos away some insects. It turns out that she's the daughter of a wealthy businessman and he is looking for a man like David to wed her to. If woke up at that moment, his future would change completely, but he kept sleeping. She watches him for a while, breathing heavily, her face flushed red. Eventually, she, too, moves on.

Next come two bandits. They look covetously at the bundle under David's head. As they're taking out knives, a big dog runs up and starts barking. They don't want the owner of the dog to discover them, so they put their knives away and walk off as casually as possible.

"In a few hours they had forgotten the whole affair," Hawthorne writes of the bandits, "nor once imagined that the recording angel had written down the crime of murder against their souls, in letters as durable as eternity." A short time later, David wakes up. In a short period of time, he had come so close to wealth, love, and death, without even realizing it. He gets on the carriage to Boston and his life returns to its regular trajectory.

---

[5] I am talking about Hawthorne's 1837 novel, *David Swan*.

If there was a record, everything would change. David could disembark from his carriage in Boston and find the wealthy couple that needed an heir, or track down the woman that seemed to have fallen in love with him, or bring the bandits to justice... But when our eyes close, we lose our insight into the world around us. And even if we do have our eyes open, it's difficult to know the true thoughts of other people. The psychological world is still a space that we can't penetrate with records. But novelists can unwind emotions, speculate, and depict the inner world of characters. Literature can allow for incredible things.

It's because of this ability to depict what can't be recorded that literature has developed the way it has. The core of literature is human nature. It's difficult to shine a light into the darkness of the human soul. It is as vast and profound as the starry sky. But literature can show us human nature. That's a unique characteristic of literature that sets it apart from science.

When parts of human nature become recordable, they can be incorporated into science, reducing the foothold of literature. This is also the reason why universal recording can change human nature.

If we could record everything and gain insight into it, our spirits would be overwhelmed. As Hawthorne says in the novel, if his protagonist could see all the possible routes his life could take, he would be filled with conflicting emotions. This also shows us that the lives of individuals are unpredictable.

So, is it better to record truth and facts and know what's going on, or should we avoid it? Transparency and happiness are not necessarily linked. It's hard to say whether or not it's better to be a prophet or a fool. This is the place where human nature and the world are entangled. If we truly understood the amount unrecordable in our human world, it would shake our faith in data.

There are things that simply cannot be recorded; there are also limitations to the extent to which other things can be recorded. As soon as data is generated, it begins to fragment. Data is like shards of a broken window pane. It is always incomplete and discontinuous. Data can only record a single fact or several facets of the same fact. No matter how the record is made, data can never reproduce the whole.

Twelve years ago, when I arrived in the United States as a graduate student, my professor repeated a famous line: "Everyone can have their own opinions, but they can't have their own facts." We must differentiate opinions from facts. From that day forward, I took the quote as a standard.

Even so, I have learned from experience that the reason why consensus is so hard to reach is because so many people do have their own facts. There might be only one objective truth, but there are countless interpretations. When people look at an event or issue, they see only one side of it. It's rare for someone to have a complete or even multifaceted understanding. So, everyone does have their own facts.

That demonstrates a huge risk in our present time: collecting a large amount of data will lead to "everyone's voice count; everyone has their own understanding." I mean that if a person wants to come to a certain conclusion, they will be able to find data to support it, even if it's an opinion that's divergent from the mainstream. No matter how much data we possess, we still may not be able to arrive at a complete understanding.

The data might come close to the truth, but even as it does, it never fully represents the truth.

If there is a God, then only He can see the complete picture. We can't come close. No matter how much data we have, we simply can't do it.

Leaving aside the issue of fragmentation, there are other reasons why our records lag behind reality. In 2016, I gave a speech at an annual meeting held by computer storage manufacturer Seagate. On my way back to Hangzhou from Chengdu, I noticed a message from an attendee in my email account. It had a rather sharp question. "Mr. Tu," it began, "I am a senior data analyst, but the rise in the amount of data we're collecting has turned me into a skeptic. I didn't want to raise this question at the meeting because I was worried how my colleagues would react...".

His issue was that he could see that more and more data was not bringing us any closer to the truth. The truth couldn't be discovered through data. In other words, even if the data was accurate, it didn't necessarily correspond to the facts.

His clarity moved me. Data records facts but, once recorded, they immediately become something from the past. Recording is always in the process of attempting to catch up with facts. This is one of the limitations of recording. The reason is that the facts are constantly changing. The value of records is always relative. I call this the theory of data relativity. Einstein's theory of relativity covers space–time and gravity, and the theory of data relativity relates to data and facts. I can't help but think of the tortuous, complicated two century process that Americans went through to figure out the population of the country:

> People within the American government have racked their brains to figure out how many people actually live there. Beginning in the 1860s, successive administrations began sending out letters to citizens, entreating them not to conceal the number of members in their family from census takers. Signed by the president, these letters promised that the information collected would only be used to arrive at true population numbers. They vowed that numbers wouldn't be used for taxation, military conscription, or court cases. Since then, presidential administrations have endeavored to eliminate human error from their results and ensure the objectivity of census data. They also did their best to speed up the census. The first census took two years, which was eventually shortened to months, and then days.
>
> At every moment, babies are being born and people are dying. These events happen in homes, hospitals, and even in the wilderness. Reality does not wait for a true portrait to be painted. In any man made census process, there is no way to gather all these facts simultaneously. They can only give us a snapshot of a certain moment.
>
> To this day, even with the advent of the Internet and mobile phones, the problem remains unsolved.
>
> Man is the wisest of all creatures, but we are still not capable of adding up how many of our kind populate this planet. What about things? They are even more innumerable. A single red bean, a bowl of beef noodle soup, a car, an emotion, a thought... We cannot even imagine. It recalls the saying, "The universe exists in every flower; the Buddha lives in every leaf."
>
> The world is large and full of innumerable things. Everything is constantly in motion. In a moment, everything can change. Like the ancients, we are only capable of taking a decision without knowing every fact, only possessing a small piece of the puzzle, a small lens to peer through on our "small facts." Our feelings of confusion and perplexity are natural and hard to overcome.

In a complex and constantly changing world, we can only do our best. Even though the population numbers could never be fixed, the American government kept attempting to improve their data collection methods, increase the efficiency of their programs, and try to get one step closer to the truth. Now, the Census Bureau has a Population Clock that can predict by-the-minute changes in the population.

Like a pendulum that never stops, data is always catching up with facts. With the improvement in recording technology, invoking the theory of data relativity, we have entered an era of improved clarity.[6]

Data is the foundation of AI. It's AI, rather than human intelligence, that will rule the data space.

Since recording has limitations and data spaces have boundaries, AI must also have limits. To understand AI, we must know where its limits are.

8.6   Facial Expression Analysis: The Boundaries of AI

In addition to playing chess, AI has three iconic applications: unmanned vehicles, speech recognition, and image recognition. We can see clear boundaries in each use case.

Let's look at unmanned vehicles first. As we have seen in an earlier chapter, although great progress has been made over the past couple years, there is still a large gap between present development and universal application. The key to developing AI for driving will be creating the types of roads and infrastructure that allow it to operate. This is not the kind of job that will be completed overnight, but it will soon be realized in certain applications, like fixed routes inside industrial ones, piloting street cleaning equipment in predawn hours, and short distance transit. Those are coming soon.

Driverless cars will result in some drivers being made redundant but I don't believe the fear of AI-driven unemployment will be durable. AI will eliminate some jobs; it will create new jobs. What we want to usher in is a new age of growth, just like the spread of machines at the beginning of the Industrial Revolution. At that time, there was fear of machinery replacing people at their jobs, but what actually happened was an overall improvement in standards of living.

In this process, we will have to pay a price, too. In an era of AI, all innovation and progress is driving toward giving intelligence to machines and bringing us closer to them. There will be many instances where we are forced to adjust our work practices, behaviors, and skill sets to meet the needs of machines. This is how man and machine will be united. At that point, we can begin to have humans and machines cooperate and complement each other. We can co-exist with machines. That will mean undergoing a degree of systematization.

---

[6] Tu (2016).

Now, let's look at speech recognition, part of the field of machine hearing. Deep learning has made great progress in speech recognition and a large number of automatic transcription and translation projects have been launched. According to Liu Qingfeng, the chairman of iFlytek, the accuracy of its transcription software has reached 97%. The error rate of machine hearing in the WSJ eval92 Benchmark has dropped to 3.1%, which is lower than the 5% human error rate. In a smaller domestic benchmark test, the error rate was 3.7%, while a 4% human error rate was found. That means that AI speech recognition has surpassed our own abilities.[7]

We can see in speech recognition the convergence of human and machine intelligence. There are times when we must modify our voices for the machine to better track them. This is an example of us being trained, or becoming more like machines. An analogy might be a field being tilled by farm equipment and then beginning to grow in a way that makes harvesting easier for the machine.

Apart from speech recognition, the field of machine hearing also involves being able to tell a person's emotional state through their voice, as well as speech synthesis, which will allow AI to speak. For sounds other than the speech, such as music, machine hearing has other recognitions, like recognizing melodies or chord progression or genre classification.

Finally, we turn to image recognition, which is part of the field of machine vision. Out of the three, this is the sphere that deserves the most attention. As we have already seen, most of the information we take in about the world around us is through vision. There is more visual than auditory information. The success of unmanned vehicles will also rely heavily on machine vision and image recognition.

As we have already discussed, the development of image recognition and especially facial recognition has great value and significance. But facial recognition goes beyond simply recognizing a face. The purpose is to one day get to a point where emotions can be determined from the study of a face. Facial expression analysis and facial recognition are not the same: facial recognition involves distinguishing faces, while facial expression analysis is using the same technology to determine emotional state from facial expressions.

It is usually said that human beings have "seven emotions and six sensory pleasures," or this is sometimes summed up in the four-character phrase, "joy-anger-grief-happiness." The easiest way to see human emotion is on the face. The study of facial expressions has become the object of study in both East and West. The idea of closely observing somebody's words and expression is one part of this, as is physiognomy. *Complete Works of Wise Advisors* records this story:

> In the Spring and Autumn Period, Duke Huan of Qi discussed with Guan Zhong the matter of punishing the Prince of Wei. He went back to the palace. When Wei Ji saw Duke Huan of Qi, she went and paid obeisance, begging to be punished in place of the Prince. Duke Huan of Qi asked her what her reasons were. She answered: "When I saw Your Highness coming, you walked boldly and I could see you had the will to punish Wei. When Your Highness saw me, his temperament changed." The next day, Duke Huan of Qi went back to the court and Guan Zhong asked him, "Have you reconsidered attacking Wei?" Duke Huan of Qi asked

---

[7] Amodei et al. (2015).

how he knew. Guan Zhong replied that he had noticed that Duke Huan of Qi was walking tentatively and speaking slowly. When he met the minister, he seemed ashamed.[8]

It's because the people around him are so good at observing his body language that they can figure out what Duke Huan of Qi is thinking. This shows us the wisdom of traditional China. They might be thought of as disloyal officials, but close observation of body language and currying favor and gaining knowledge through it was a necessary skill. Like facial recognition for AI, it requires a lot of data and complex calculations. Disloyal officials could use their past experience with the sovereign to make judgements. They had minds that were adept at those calculations. They were intelligent, even if they were dishonest.

Digression: the historical popularity of physiognomy

Looking for patterns to decipher in facial features was the earliest form of facial expression analysis. As I have already mentioned, a craze for portraits followed the invention of photography, and following that was a rise in interest in studying faces. In East and West, there has been an interest since ancient times in the connection between the human face, individual characteristics, and destiny. There are many idioms in Chinese that imply this connection, like, "the face follows the soul," or, "our faces are our fate." The former suggests that our emotional state will be written in our appearance, and the latter suggests that the face reflects some inborn personal characteristics that will guide our destiny.

Physiognomy was once broadly popular. In a 1927 murder case in the United States physiognomy experts were called to testify. Ruth Snyder was charged with attempting to kill her husband seven times and succeeding on the eighth attempt. The expert took measurements of Snyder's face with calipers. He noted the pointed, feline character of her chin, which he said denoted a treacherous and ungrateful individual. Her facial features suggested had "the character of a shallow-brained pleasure seeker... murderous passion and lust."[9] Snyder was sent to the electric chair in no small part because of this testimony.

We now know that physiognomy is pseudoscience, but that has not extinguished its vitality. There are still experts that claim it to be genuine.

Like many other sciences that grew out of primitive practices, facial expression analysis grew out of physiognomy in the 1970s. It overlapped slightly with physiognomy and the idea of studying someone's words and expressions to learn their true thoughts, but it surpassed these to become a new science.

---

[8] This is drawn from the Ming Dynasty work Complete Works of Wise Advisors by Feng Menglong.
[9] Sacks (2003).

The founder of this new science was Paul Ekman (1934–). He and his colleagues spent eight years creating a reliable method to analyze facial expressions. They began by breaking down the anatomy and identifying 43 facial muscles. Each muscle was a unit of action. All facial expressions could be understood as a combination of these 43 units. These combinations formed a facial expression code system.

This system was made up of a database that uncluded 10,000 photographic and text descriptions of facial action unit combinations. Each piece of data describes muscles, unit combinations, and their corresponding emotions. Ekman believed that there were 3,000 facial expressions relevant to emotion. Those facial expressions were interpretable. Ekman began experimenting on himself, mobilizing every muscle in his face to make a particular expression. When he failed, he went to consult with a surgeon, who helped him use a needle to stimulate those particular muscles that remained uncooperative.

This database was very effective. With it, Ekman wrote many legendary new chapters in the history of psychology.

Suicides are common in psychiatric hospitals. Patients that want to kill themselves will often go to the doctor and report, "I'm feeling fine. Can I go out for a stroll?" Experienced doctors know that patients will sometimes say this sincerely, but the second possibility is that they are looking for a chance to commit suicide. It is difficult for doctors to make that call (Fig. 8.2).

Ekman was once ranked among the top hundred psychologists of the twentieth century. He has trained tens of thousands of lie detectors across many fields. He has found that the most successful group was people from intelligence and special services due to their experience with studying facial expressions. He found that judges were often the least effective at judging the sincerity of someone testifying, since they sit in a place where they can't see their faces and are often absorbed in notetaking or listening. Ekman's work and practice provided the material for the celebrated TV series Lie to Me.

Ekman asked the doctor to record the process of their conversation with patients. He then watched them repeatedly. At first, Ekman didn't notice any difference between the sincere and the suicidal patients. However, once he reviewed them

Fig. 8.2 American
psychologist Paul Ekman

Fig. 8.3 Screenshot of the White House press conference that included Clinton's denial

multiple times in slow motion, he could see the difference: the suicidal patients in expressions that lasted less than 0.07 s exhibited strong pain. These expressions were short but revealed the patient's true intentions. Ekman later identified them in other footage. He called these "micro-expressions." These brief expressions that flashed across people's faces could not be detected by untrained observers, but they revealed the true intentions and feelings of the subject.

Apart from those pained micro-expressions, Ekman was also able to tell from his system of facial expression analysis whether or not a person's smile was genuine. A genuine smile is caused by spontaneous emotions. It mobilizes the muscles around the cheekbones and eyes that are incapable of being moved consciously. The forced smile, though, uses the zygomaticus major, which extends from the cheekbones to the corners of the mouth. Another example was the frontalis pars medalis, located in the area of the inner eyebrow. When slightly raised, it reveals sadness. "If you see this, you'll know the person is sad even before they know it."[10]

We must remember that the foundation of Ekman's micro-expressions was recording. The video footage of patients interacting with doctors was crucial material for Ekman's research. He used video instead of photos because there was less chance of the subject striking a pose and obscuring their natural state. When researchers try to capture changes in facial expressions, photos are not sufficient. This also illustrates the reason for the emergence of film and cameras.

In 1998, Bill Clinton was suddenly embroiled in a sex scandal involving a White House intern named Monica Lewinsky. At first, Clinton denied having had sex with Lewinsky. Clinton claimed this: "I did not have sexual relations with that woman." When Ekman saw it, he knew Clinton was lying: "One of the things people do when they lie is to use distancing language," Ekman said. "We know he knew Monica. Yet he used the words 'that woman.'"[11] He was even capable of pointing out exactly which facial muscles Clinton had been using (Fig. 8.3).

---

[10] Guthrie (2002).

[11] Ibid.

Psychologists quickly determined that Clinton lied at the press conference. Psychologists have found that touching the nose is a sign of deception. The reason for this is that lying increases heart rate, which makes your nose twitch. Clinton touched his nose 26 times during the press conference.

Once Ekman's system was proven effective, there were proposals to train AI to make use of it. Ekman himself predicted in 2004 that his system would be usable by AI within five years and that there would be cameras capable of immediately processing micro-expressions.[12]

Digression: Other uses of facial expression analysis

Facial expression analysis has a wide range of uses beyond lie detection. In business negotiations, being able to interpret subtle changes in expression would afford a great advantage. In sales, careful observation while a customer inspects a product could give hints as to their true thoughts. When it comes to individual use, it could give people a better understanding of their own facial expressions, which will usually only be seen by their family and friends. Facial expression analysis could help people control their emotions and improve their communication skills.

If machines are capable of facial expression analysis, they can begin to understand human emotions and thoughts. That will be a great step forward in the development of AI. Machines will understand something of human nature. That is an important way to expand the horizons of human–machine interaction.

This might sound far fetched, but it's quite logical. Ekman has taken the inborn ability of humans to read facial expressions and generated rules for them that can be followed by machines. With facial expressions systematized, machines can understand and mimic us. In fact, AI has a great advantage over us, when it comes to interpreting facial expressions. As Ekman says, the micro-expression is usually too quick for the naked eye to perceive, but cameras have no such limitations. They can capture micro-expressions quickly and accurately. With that ability, it seems clear that AI will one day surpass the average person in being able to read faces—and might one day even be more accurate than the treacherous officials in ancient records.

Since 2010, there have been various attempts around the world, using Ekman's system as their basis, to deal with the problem of AI facial expression analysis. One example is Computer Expression Recognition Toolbox (CERT), developed by the University of California, San Diego. CERT can automatically detect faces in video footage and simultaneously recognize thirty action unit combinations from Ekman's system, including anger, disgust, fear, happiness, sadness, surprise, and contempt. A joint test between Carnegie-Mellon and the Massachusetts Institute of Technology (MIT) found an accuracy rate of 80.6% for CERT.[13] The system could be used to

---

[12] Face to Face: The Science of Reading Faces: Paul Ekman (2014).
[13] Littlewort et al. (2011).

check for signs of depression, schizophrenia, autism, anxiety, and other disorders, but it could also be installed in a car to monitor a driver's fatigue, or installed in an elderly care facility. Most people will not give voice to certain sentiments, but their facial expressions will always reveal their true feelings.

In May of 2018, Hangzhou Number 11 Middle School introduced an "Intelligent Classroom Management System," which scans a surveillance photo of the classroom every thirty seconds and identifies emotions of happiness, sadness, anger, or disgust from facial expressions, as well as common classroom behaviors (raising hands, writing, standing up, listening, putting head down on desk). The system can use this to provide statistical analysis to teachers for classroom management.

When the story of the Hangzhou class monitoring system made the news, it aroused some controversy and debate. Supporters insisted that the system would improve classroom order and optimize the learning experience, while opponents claimed that it was a violation of privacy. That violation of privacy from a young age would distort the behavior or students. It was not, opponents said, optimization but alienation. It was twisting education.

I don't believe this opposition is really warranted. A teacher's attention is limited to a small number of students at any given moment, so they will not notice other kids drifting off, fidgeting, or even leaving the classroom. The intelligent monitoring system is capable of seeing the classroom from six angles. It can fill in the gaps. There is no question that it should be adopted for use by schools. The question, rather, is how exactly it should be used.

Students can be distracted, but the machine cannot be distracted. Teachers and administrators have to take a relaxed attitude toward that kind of behavior. Immediately scolding any student that drifts off is not acceptable. That is how you turn humans into robots. In the future, the solution might be this: when the system notices that a student is distracted, it will send a signal to their smart watch, signaling them with a vibration to pay attention.

Going beyond this story, we must acknowledge that we will have to get used to being observed by machines. We will have to live in harmony with algorithms. From a young age, our children need to be aware of data force. Being surrounded by cameras must be no more strange than being in nature.

Looking back over history, the Agricultural Revolution and the Industrial Revolution have brought profound, liberatory change to our material lives, and the AI revolution taking place now will have equally significant liberatory change. It will change our spiritual, cultural, and religious lives. It is not going too far to say that it might change human nature itself.

Systems similar to the one in the Hangzhou classroom are being employed in movie theaters. The goal is to accurately gauge audience reaction. Disney designed a system to analyze audience facial expression: factorized variational autoencoders (FVAEs). They instrumented a 400 seat movie theater using four infrared cameras. In a completely darkened theater, the system can detect laughter, a slight smile, tears, and other emotional responses. By analyzing these facial expressions, Disney can get data on audience response to the movie and to individual scenes. This is a method of getting quantitative data on movies (Fig. 8.4).

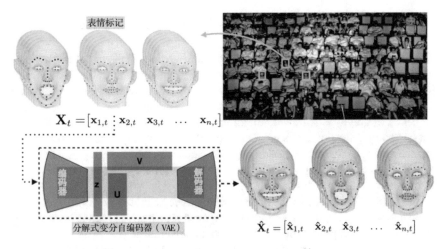

Fig. 8.4 Disney's model for analyzing audience facial expressions[14]

This is also an example of data and AI entering the artistic sphere. We used to be under the impression that machines could only deal with scientific issues and that art was beyond their abilities. There seemed to be a clear boundary between the two spheres of art and science. But science has broken down that boundary. Logic and rules can apply to art, too. There are parts of art that can be thought of as science.

My prediction is that facial expression analysis and emotional computing will be helped along by data gathered through a wider range of sensors, including wearable devices. Physiological data from human subjects will be collected, including facial expressions, speech tone, gestures, brain waves, and cardiovascular state. That will allow for a comprehensive analysis of a person's emotional and physiological state. Machines are more capable than humans in making interpretations and predictions from this data. At first, this will be restricted to laboratories. This also seems to be proof that an era of universal records will be accompanied by universal calculation.

The AI is capable of using 43 different muscle combinations to recreate human facial expressions, while Ekman's system includes more than 3,000 examples of meaningful human expressions. When we get right down to it, once rules for something are established, AI can begin to replicate it: in the future, machines will be able to have facial expressions, just like we do.

At this point, we reach one of the boundaries—if we can summarize things through logic, rules, and data, AI can begin to approach human behavior, but there are also things that are not expressed according to rules. At that point, AI is powerless.

Reading about the legendary academic exploits of Ekman is thrilling. We can marvel at his understanding of human psychology. It's an amazing feat to turn facial expression analysis into hard science. However, when we learn that machines will soon possess the ability to mimic our faces, we are filled with fear and disgust. It sends a shiver down our spines. It's undeniable that this future is on the horizon.

---

[14] Deng et al. (2017).

8.7   The Smart Society Transformation

AI is propelling us into a new social form: the smart society. In the social environment of the smart society, we will reach the peak of granularity, individualization, and integration of machine intelligence. But new crises wait around every corner. This is especially true at the current stage, when we have one foot in the smart society and our center of gravity is still in the traditional society. This is a stage of development that has many potential problems. Our society is like an exquisite crystal ball or a porcelain vase: the more delicate and sophisticated it is, the more fragile. This is the danger we face in entering the smart society. We must deal with the combination of sophistication and fragility.

The reason for this is: in the smart society, there will be a powerful process of social networking. Nothing like this has happened previously in human history. This will also have an unprecedented effect on social governance.

In the past, society was mechanical, inelegant, and loose. It was like a large clock. A clock like that is not only large and cumbersome but also tends to lose a few minutes a day. But the smart society will be digital and electronic. It will be driven by computer chips. Instead of a large clock, it will be like a massive assemblage of individual quartz watches. These units will cooperate with each other and make decisions together. These small units are far more elegant and precise than the big clock. They are also more flexible.

In agricultural societies, time is structured by the four seasons, harvest in autumn and store grain for the winter, human lives pass through all their stages… The structure of time was simple. If a hoe is broken, it can be replaced in time for planting. A farmer's problems rarely spread out from the individual to the community.

In the industrial society, machines sped up society and increased the complexity of its structures. A problem with a single part of a machine would cause it to grind to a halt. A small problem can quickly cause the paralysis of an entire system. An entire city could be plunged into darkness by the failure of a small component. That was unthinkable in the prior age of candles and lamps. In 1986, when NASA's Challenger broke into pieces 73 s after launch, killing seven astronauts, the cause of the accident was found to be record low temperatures reducing the elasticity of rubber O-rings. The failure of the O-ring allowed the unexpected release of hot, high-pressure gas, causing an explosion. This proves that no part is truly redundant and no detail can be left to chance. Each country has to invest resources in the type of management that will prevent accidents.

In the smart society, everything will be stitched together by software, the Internet, and sensors. This is a great stride forward from the foundation of industrial society. Compared to rockets and shuttles, the level of complexity will be even greater. Human error can be isolated; we can troubleshoot, repair, and improve glitches in machines or software; and in a society dominated by intelligent linkages, there will be no dependence on a single individual or component but rather many systems of increasing complexity. So, in a smart network with various sensors, controls, algorithmic protocols, and databases, there will be a high level of interdependence, so a problem within one parameter might cause a chain reaction through the system, freezing a large sector. Troubleshooting a problem in such a system will be difficult. It would be a disaster if the only option was to shut down the entire system.

There's a simple explanation for this. Networking everything will allow the butterfly effect—a butterfly flaps its wings in Brazil and causes a tornado in Texas—to be even more widespread. When we try to figure out where the tornado came from, it's very difficult to trace it back to the butterfly in Brazil.

A more complex society means that correcting issues becomes more complicated and costly.

The uneasy truth is that there will be vulnerabilities and they will appear without warning. It's been 33 years since the first Windows release in 1985 and it has required continuous patching and upgrading over this time. In 2015 alone, Windows announced 135 bugs and vulnerabilities were fixed. These vulnerabilities could have been exploited at any time. The defense has to keep pace with the offense. It's like the old saying about the shield that's invulnerable and the spear that can pierce any material. All software and mobile applications are faced with this kind of contradiction.

Murphy's Law says that if anything can go wrong, it will.

Sometimes failures can be harmless, but they can also be disastrous. On July 20th, 2016, Southwest Airlines was crippled by a failed router. 2,300 flights were canceled over four days, stranding thousands of passengers. Southwest's CEO explained it like this in a memo: "When the router failed, the data … piled up like a freeway traffic jam."

Digression: Why sensors are the key vulnerability of the smart society

According to Juniper Research, the number of IoT devices will reach 38.5 billion by 2020, a 285% increase over the 2015 number of 13.4 billion.[15] Networked sensor devices will be present in every corner of our world and become a key component of our society. These IoT devices are vulnerable to attack because of their limited capacity for running security software.

In October of 2016, the American firm Dyn, which provides Domain Name System (DNS) services to numerous web platforms suffered a large-scale Denial-of-Service (DoS) attack. The DoS attack on DNS services took down services in multiple cities, causing estimated losses to web companies totaling hundreds of millions of dollars. The IP addresses of the machines used to coordinate the mass DoS attack showed that they were mostly IoT devices. In other words, hackers had harnessed the networking capabilities of devices as mundane as Internet-connected coffee machines. In 2014, hackers hijacked 100,000 IoT devices, including fridges, routers, and smart TVs to launch attacks.

---

[15] Internet of Things Connected Devices to Almost Triple to Over 38 Billion Units by 2020 (2015).

There was once an incident where the Thai finance minister Suchart Jaovisidha was trapped in his BMW after the onboard computer malfunctioned. The air conditioner was switched off and the doors and windows were sealed shut. Onlookers were powerless to help him. Finally, a security guard from a nearby building arrived with a sledgehammer and broke a window.

As the complexity and intelligence of a system grows, the possibility of fatal glitches actually increases. This risk shows us the fragility of society.

Humans are prone to error, but in the smart society, the mistakes will come from computers and algorithms. That might be even more disastrous. My prediction is that we will soon realize that algorithms and networking can only expand to a certain point. There will soon be legislation to restrict algorithms. Our ultimate goal has to be finding a balance between human and algorithmic autonomy.

We can compare our new smart society to a beautiful vase of flowers. The vase represents its fragility and the flowers represent a world of sensuality. In the future, we will have a world with many beautiful flowers of information, entrancing our gazes. Mixed among them might be fake plastic flowers. As machine vision and machine hearing improves, as well as facial expression mimicry and machine speech, it is more likely that these technologies will be used to spread falsehood.

In March of 2016, the University of Nuremberg released an application called Face2Face, which allows for real-time face capture of a person that is then projected on another face. In other words, you can record yourself saying a phrase and then produce a video of someone else saying the same phrase in the same way. Face2Face is not the only program capable of this real time "facial transplant," but it's particularly accurate and realistic. The results are hard for the untrained eye to differentiate from real videos.

In August of 2017, researchers at the University of Washington published a paper about a project that involved using a neural network and machine learning to produce a deep fake of Barack Obama delivering a speech. The program had analyzed millions of frames of video to flawlessly copy Obama's face when he spoke. It was capable of making Obama say things he had never said before.[16] In March of 2018, iFLYTEK CEO Liu Qingfeng said in an interview that the dream of his company was to make machines speak. He claimed that they would one day be capable of imitating Trump so flawlessly that even Americans would believe it.[17]

This technology can be used to mimic the speech of any world leader, public figure, celebrity, or private citizen. It could be used to make a fake speech from Trump or Kim Jong-un. Imagine for a moment the implications of this. Even a brief speech about nuclear weapons negotiations or testing could have a massive impact on the stock market (Fig. 8.5).

At upper left is the person providing their expressions. The software copies the expressions and projects them on the target object, as well as synthesizing their voice. Here, we see a synthetic George Bush, Jr. mimicking the expressions of the person at upper left.

---

[16] Langston (2017).

[17] Hu (2018).

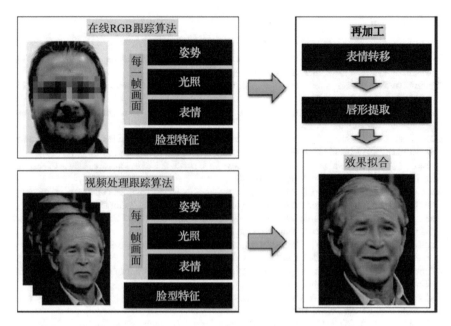

Fig. 8.5 Real time facial transplant by Face 2 Face, University of Nuremberg[18]

When faced with these deep fakes videos, images, or audio, it is difficult for even professionals to quickly verify them. To combat fraudulent behavior, we need to improve machine learning. In the future, AI might be able to detect deep fakes. They will be able to notice the lack of particular flaws or some other characteristics that a real video would have. As we enter a smart society, there will be an arms race between fraudsters and anti-fraud agencies.

This new arms race is concerning. When machines become capable of sensing, remembering, learning, analyzing, making decisions, and mimicking, they will quickly reach a point where they are millions of times more advanced than ourselves. The possibility of a negative outcome is magnified. Technology is neutral; people with good and bad intentions will endure the outcome of technological advancement. Data space is virtual, but we don't want it to be completely fictional. The problem is that history has repeatedly proved that human nature lacks the ability to judge or resist change. The paradox is that only AI can protect us from AI. This is the key contradiction of the era. We have to charge forward. There is no going back.

---

[18] Thies et al (2016).

References

"Face to Face: The Science of Reading Faces: Paul Ekman." (2014, January 14). Conversations with History, Institute of International Studies, UC Berkeley.

"Internet of Things Connected Devices to Almost Triple to Over 38 Billion Units by 2020." (2015, July 28). Juniper Research.

Amodei, D., et al. (2015). "Deep Speech 2: End-to-End Speech Recognition in English and Mandarin." arXivLabs.

Deng, Z. et al. (2017, July). "Factorized Variational Autoencoders for Modeling Audience Reactions to Movies." 2017 IEEE Conference on Computer Vision and Pattern Recognition (CVPR).

Guthrie, J. (2002, September 16). "The lie detective / S.F psychologist has made a science of reading facial expressions." *San Francisco Chronicle*.

Hu Chunyan. (2018, March 11). "Liu Qingfeng: 'We will have machines capable of mimicking Trump so perfectly that even an American would fall for it." China Youth Online.

Jack Ma. (2017, May 26). "Complete text of speech at China International Big Data Expo: The next 30 years will be an era of overtaking and redefining." Sina.

Chen Jie. (2016, November 21). "Jack Ma conference speech: the 'planned economy' will continue to expand over next three decades." *Qianjiang Evening News*.

Langston, J. (2017, July 11). "Lip-syncing Obama: New Tools Turn Audio Clips into Realistic Video." University of Washington.

Laplace. P., translated by Truscott, F. W. and Emory, F. L. (1951). *A Philosophical Essay on Probabilities*. Dover Publications.

Littlewort, G., et al. (2011, March). "The Computer Expression Recognition Toolbox." Ninth IEEE International Conference on Automatic Face and Gesture Recognition.

Sacks, O. (2003, April 27). "Would you lie to me?" *Guardian*.

Thies, J., et al. (2016). "Face2Face: Real-Time Face Capture and Reenactment of RGB Videos." IEEE Conference on Computer Vision and Pattern Recognition (CVPR).

Tu Zipei. (2016, March 31). "The Era of Big Data: A new theory of relativity for a new age." The Paper.

Epilog: The Fourth Wave and How We Retake the Lead

The year 1500 AD is regarded by many scholars as the beginning of the modern world. American scholar Leften Stavros Stavrianos (1913–2004) uses that year as a demarcation point in his A Global History because it was when the world began to be integrated together, including by the great age of exploration that linked Europe, Africa, and the Americas.

Before this period, China had been a great power, but it began a gradual decline. Many reasons have been proposed for this.

Looking back at the case of recording, I believe the main reason for China's decline is that the printing press was invented in Europe. We can refer to that technology as the Third Wave, following the invention of writing and paper. China missed this wave. That is one of the remarkable things about human history: when an opportunity arises, those who seize it will open themselves up to more opportunity; those who fail to seize the opportunity will find other opportunities closed. The latter loses their original advantage. This is called the Matthew effect of accumulated advantage. That was a problem for China until the second half of the twentieth century.

After the invention of the printing press came the camera in the 1840s and then the audio recording equipment in the 1980s. We were suddenly capable of keeping images and reproducing sounds. These were unbelievable advances. With these new tools, we could expand recording from simply noting things in text. But the equipment was expensive and not portable, so people had to go to specialized venues for photography or recording. These new technologies remained in the hands of professionals, at first.

Over the final decade of the twentieth century, things began to change. The rise of the Internet allowed the unprecedented spread of recording. Information such as text, audio, and images began to be digitized. Apple launched their first iPhone in 2007 and smartphones began to appear in the palms of consumers, ushering in an age of universal recording. This is the Fourth Wave of recording, following paper and text, then the printing press. Almost everything in our lives can now be recorded. This data can be stored in the cloud and be passed onto future generations.

Looking back at the sweep of history, human beings at the start of agricultural society made their living from the Earth and survived on what it produced, but they

© China Translation & Publishing House 2022
Z. Tu, *The New Civilization Upon Data*,
https://doi.org/10.1007/978-981-19-3081-2

had little ability to impact it. Once we entered the industrial age, we erected massive structures on the Earth and began remaking its shape, molding and creating new materials, building out of glass, steel, and cement, and creating infrastructure for water, electricity, and gas. The city became the center of human civilization and gathered the elite of our species. On the surface, we can analyze cities as a semi-artificial entity. In the age of information, when data has become a new resource, linked together by the infrastructure of the Internet, we have a new bedrock to deposit information. Data has increased in quantity and has been further integrated. That has allowed us to create a new space—the data space. There is now a virtual space beyond the physical world. In this new space, data is like a new soil that can grow all things. Humanity is its master.

In other words, the data space is a completely man made world.

We have received some blessings from this new virtual space. Compared to the twentieth century, the individualization of universal recording gives the individual new potential. The data space will allow that potential to be realized. The world will be rapidly transformed by this new reality.

Today, anything we write can be uploaded and then searched, analyzed, and quantified through the store of knowledge that is the Internet. If you want to find a sentence similar to the one you're reading in another book, you can do it. You can check whether your thinking, language, and logic is novel or not. We can now stand on the shoulders of giants and take genuine leaps forward. This is innovation.

Around the Lantern Festival in 2007, I talked to a poet in Hangzhou named Wang Ziliang and he gave me a book of his poetry. He told me that he had always wanted to write a poem about cats but had avoided the topic because he felt Baudelaire had already mastered it in his poem called "Cats." It got me thinking that a good poem is one that cannot be rewritten. The same is true of a good essay or book. If it cannot be replaced, it becomes the "boundary."

Human knowledge has boundaries but knowledge itself has no boundaries. If you want to be a pioneer, you must know where the boundaries lie. To move forward, you must know where you stand and where your predecessors have already gone. Innovation is easier when it proceeds from the work of people that have gone before.

If we compare all existing knowledge created by ourselves and our predecessors to a circle, then at any point on the line, people may have made breakthroughs. That expands the circle. The limits of human knowledge can be expanded.

Here, it must be said that most previous expansions of that circle of human knowledge were made by professionals. This is because finding the limits and building on the advances of predecessors requires some expertise. The cost of innovation was formerly quite high, but the reality of data and Internet expansion, as well as real-time, inexpensive, and universal records, means that barriers have been lowered. People outside of professional spheres can now offer their own contributions to innovations. In the past, innovations by ordinary people existed but were rare; in the future, that will change.

Now that all professions are being affected by data, we see the creation of large-scale database records. This is not only to store data but also to integrate and optimize it. There are certain problems that could not be easily solved if our level of technology had stayed at the printing press. Once a book is published, the technology of the printing press makes it more durable, but it cannot be updated. Information recorded online can be continuously read but also edited and discussed. Examples of this are Wikipedia articles or Dianping entries. There is a cycle that means records on the Internet are dynamic, constantly changing, and also linked. Finding information online and finding information in books is completely different.

A large proportion of what's recorded online are individual activities or personal opinions. This information can be integrated to produce a new kind of subjective authority—the wisdom of the crowd.

In human history, organizations, corporations, and specialists used to take advantage of the fact that certain things were unrecordable and unquantifiable. They could use this to manipulate people. But because this ability has been diminished because of universal recording. Now, experts are being replaced by popular consensus. This consensus now often takes the form of data. For example, when we shop for a product, we can see its sales figures, reviews, and how many people have hit the "Like" button on it. This can have a direct impact on our behavior as consumers.

Data is shining a new light on our world. This light is similar to the natural light emitted by stars; it appears weak to us, but in fact is produced by astral bodies just as powerful as our own sun. The closest star to the earth is Proxima Centauri, which is four light years away. In other words, traveling at the speed of light, it would take four years to reach it. It's more than 27,000 times further away from the Earth than our own sun. The only reason it seems so dim in the night sky is because of its distance. If we were to get close to it, we could appreciate the massive amount of energy it releases. The same holds true for data: it seems distant and dim, but it also holds an immense amount of power that we can only appreciate when we get close to it.

In the long term, I am optimistic about the development of Chinese civilization. Over the past three decades, we have seized on the opportunities offered by the Internet for building social infrastructure. Europe might be the birthplace of the printing press and the Industrial Revolution, but they have failed to seize on the Internet revolution, just as China failed to get onboard with those earlier innovations. Up until the present, Europe has not produced any globally competitive Internet firms. Instead of seizing the opportunity, Europe has fallen behind and the glories it decorated itself with during previous ages have now dimmed.

The Internet emerged in the 1990s, and in the first year of that decade, the American government lifted most restrictions on its consumer use. That allowed the internet to go from a specialized network for military and industrial research to a commercial public network. Since then, innumerable companies and individuals from around the world have jumped in headfirst. China made sure that they would be part of the Internet—and they have even gone beyond participating, now, taking a leading role in some sectors.

In 2005, the number of Internet users in China broke the hundred million mark, and, by 2017, that number had reached 772 million (753 million mobile users).[1]

---

[1] *People's Republic of China 2017 Economic and Social Development Report*. (2018, February 28). National Bureau of Statistics.

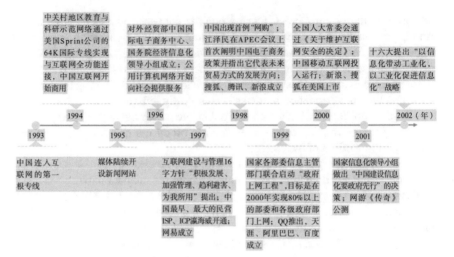

Fig. A.1 The golden age of the Chinese Internet, 1993–2002

There are five major Internet companies in the world and China produced two of them. China has boarded a bullet train to the Internet future. There's no doubt that the country is a force to be reckoned with.

Since the reforms of the 1980s, China's leadership has embraced computer and networking technology and adopted an attitude of "develop first and manage later." For most of that period, China has not had a competent authority in charge of the Internet. That provided space for the Internet to develop (Fig. A.1).

The "develop first and manage later" approach turned out to be of great benefit. That remains true. We have had an open mind to the Internet. In hindsight, the leadership sees the wisdom of that decision. In the wake of major changes in the field, with the arrival of big data and AI, China's arms remain open.

In August of 2015, the State Council produced its "Notice of the State Council on Issuing the Action Outline for Promoting the Development of Big Data," which included establishment of a management mechanism of "talking, decision-making, management and innovation based on data." In July 2017, we saw the release of the "Notice of the State Council on Issuing the Development Plan on the New Genera-tion of Artificial Intelligence." This AI plan pointed China in the direction of taking initiative on adapting to technological and social changes brought about by machine intelligence. China does its best to take the lead in development, as it represents a historical opportunity. There is a three step goal for AI in China: reach parity with global AI development and application by 2020, begin achieving major break-throughs and setting the pace by 2025, and turn China into a world leader and the center of AI innovation by 2030.

The plan for AI in China contains many advanced, eye-catching, and unexpected statements, especially in the realm of human–machine intelligence, crowd-sourced AI, smart economy, smart governance, transmedia integration, etc. I can tell from

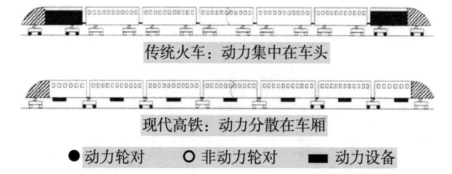

Fig. A.2 Differences between the drivetrain of a traditional train and high-speed rail

the plan that the Chinese government is ambitious. I am happy that our leaders are seizing this historic opportunity. Two centuries ago, China missed out on the Industrial Revolution, but we won't make the same mistake this time.

The world is changing quickly. Nowadays, when I take a trip, it is often by high-speed rail. This transportation innovation has become a symbol of China. It is far faster than traditional trains because of advances in its drivetrain. Everytime I board a train, I think about comparative types of advances taking place in society (Fig. A.2).

Over the long development of the technology, the locomotive was always given the role of driver of the entire length of the train. So, the speed of the train relies on the speed of the locomotive. The driving force was centralized and individual. But the weight that the locomotive was pulling limited its speed. The power of high-speed rail lies in the fact that all cars are providing drive. This is a distributed system. Rather than being dragged, the cars are driving themselves. The cars in the length of the train are in a state of absolute coordination. The is not only a mechanical and energy synergy but also an information and data collaboration. That is why high-speed rail can run so quickly.

This dynamic collaboration is similar to the architecture of cloud computing and big data. The traditional system is that information is stored in one place; when a calculation needs to be made, the information is pulled from that location. So, the computer drags a database behind it, providing the computing power. In the distributed cloud, the operation is similar to high-speed rail, with each node providing computing power. Connected to the cloud, a computer can store data, as well as performing computing tasks. The node provides storage and also processing capabilities, rather than just being a passive part of a system.

If we look at the architecture of the blockchain, we also see a distributed system. After a transaction takes place, the blockchain records information at multiple links on the chain. Every node is capable of storing information. If we compare it to a company, it would be like having every employee capable of doing accounting tasks; everyone has access to the ledger; and they are equal in power, so they can supervise each other and reconcile their records with other ledgers on the blockchain. Each node can also act as the master node, if required.

Integrating high-speed rail, cloud computing, and blockchain technology into our society will lead to significant dynamic change.

That dynamism is why I have so much faith in the future progress of Chinese society.

It seems to me that China's leaders are acutely aware of this kind of historic change. In 2015, government documents began including this slogan: "mass entrepreneurship and innovation initiative." This eventually was changed to "upgraded mass entrepreneurship and innovation initiative," reflecting the widening of the goals to activate hundreds of millions of people in the promotion of the nation's development. Looking at it another way, we could recall the distributed systems of cloud computing and high-speed rail, and see that the same logic might be applied to the people of the nation itself.

As I write this, China's Reform and Opening is entering its 41st year. In the past, that policy was driven by the state and elite actors, but now it requires all citizens to participate. The development of any country requires a sustainable driving force, and we must draw from every citizen.

Technology is empowering the individual. Our civilization is turning the page on a new chapter in its development. That will require shifting from state-driven to citizen-driven progress. This mechanism is the most fundamental guarantee of China continuing to flourish and remain as a world leader. If we can continue to develop by this means, we can usher in a new era. A country with a big population and big data is sure to have a bright future.

Afterword: Setting Out on the Rugged Trail From Which No Man Has Yet Returned

This is my third book on the same topic, following *Big Data* and *The Peaks of Data*. I tried making some changes from those works.

The first change was the perspective. The first two books were written while I was in the United States and are more about looking at China from the outside world. By the time I started work on this book, I had been back in China for three years and had gotten some experience as a corporate executive here. This book is my reflection on current trends in development and reform from within the country.

The second change was that after spending eight years mostly reading English works on the topic, I started re-reading some Chinese-language works, especially the classics. I hoped to gain inspiration from traditional Chinese wisdom. Since I am writing in Chinese, I thought it wise to also try to introduce some ancient Chinese wisdom. These works can offer solutions to problems of modern times.

One example is my writing on quantum mechanics. I was struck by two poems by Tao Yuanming. I was struck by the image he uses of human life being like dust. He goes on to say that the gust of wind itself gives form to the formless. The basic meaning is that life is like particles scattered in the air. Our bodies take on different forms through hardship, so we finally end up as something else. This inspired me to conceive of modern people as the elementary particles of the city. There is no reason not to apply quantum physics to urban problems, since people can be just as unpredictable as particles.

Another example is incorporating notes on ancient Chinese painters into my chapter on facial recognition. The Western Jin Dynasty scholar Fu Xian (239–294) wrote specifically on the issue of portraits and neatly summed up why aristocrats demanded pictures of themselves: they wanted to leave a lasting legacy. They didn't want to be forgotten by future generations. The earliest portraits came from the same place as modern portraiture, which is recording. The same holds true for East and West.

The need to record things is part of human nature.

The third change I introduced in this book is that it is more focused on the future. It attempts to answer questions on the cutting edge of technology. Due to the arrival of deep learning, artificial intelligence has made leaps and strides in the past five years,

© China Translation & Publishing House 2022
Z. Tu, *The New Civilization Upon Data*,
https://doi.org/10.1007/978-981-19-3081-2

but what is coming next in AI? Deep learning uses data as a way to develop AI, so what would happen if the machines could begin to digest books? Would that allow them to have more human intelligence? Another issue to consider is errors. Human drivers will have traffic accidents and human doctors will misdiagnose illnesses, but what about machines? What are the limits of machine intelligence? Humans also have the ability to process tacit, unspoken information that cannot be systematized or clearly put into words. That is why doctors are considered more knowledgeable as they age. They acquire and accumulate wisdom with age and experience. There are things they know that might be hard to pass on to future generations. So, the question arises, will there be a new form of civilization beyond what we see around us now?

While algorithms and AI have made great progress in approximating human intelligence, human existence is moving closer to machine life. In the 1930s, Charlie Chaplin's Modern Times gave us an image of an alienated assembly line worker screwing in screws all day. What most people don't realize is that people using their smartphones are also in the same predicament. The modern phone is also like working on an assembly line, just like all software. This assembly line remains hidden in the cloud. The alienation of industrial society was confined to factory workers, but the alienation of the data society is universal. Today, no matter where we find ourselves, the assembly line beckons, but instead of screwing in screws, we are scrolling and tapping.

I agree with Marvin Minsky that most humans are machines most of the time. So, AI is certainly capable of matching our average intelligence. In fact, it will likely surpass it soon.

But can AI be put to any good use by ordinary people? Would implanting chips in our brains allow us to quickly retrieve stored information and spit out facts like a robot? What I mean by this is, will we join the machines in surpassing our present average intelligence? Will the answer be in augmenting our intelligence? Could this work? Could this enhance human creativity?

I also tried to look at the issue of algorithms and their power to integrate data. The spread of algorithms reduces the random chance inherent in human society. That serves to make life more efficient, but not necessarily better. The integration of data provides more opportunities to stitch society together, but there are still many loopholes. In the end, will a smart society be fragile or durable?

These are the sorts of questions that emerge at the outer limits of human knowledge. I have attempted to answer some of them.

Many nights, I paced back and forth in my office or sat in front of my computer, trying to get the basic logic or facts of my arguments correct. During these times, a poem by Li Yu of the Southern Tang dynasty would come to mind. He wrote: "I see snow on the distant mountains/Leading to them, a rugged trail from which no man has yet returned."

The poem describes a man wandering alone up into snowy mountains. In these lines, we get the sense of a man that is alone and powerless to resist the call of the trail. While writing this book, I had the revelation that the ancient loner would be today's startup guy. The powerlessness to resist the rugged trail that he describes might be the reaction that drove him to innovation. In fact, the urge to create new solutions is the same as the urge to go into the mountains. We don't know what might be found there. The loner will climb with his bare hands up those sheer cliffs and keep going until he reaches territory where no man has been before. Most people could not bear that feeling. Human nature would push us to quickly retreat. Most people would turn back halfway.

But I also know that it is unimportant to press ahead too far. As long as someone can expand the limits of human knowledge a small amount, he has still contributed to human civilization. That is how it has always gone.

Attempting to answer these questions at the edge of human knowledge is also an innovation. That is what was on my mind while writing this book.

My inspiration was this: in actual practice, innovators must be good at using words and actions to break through the boundaries of human knowledge.

I also felt that the answers for some questions could not be found in science, logic, criticism, or precise arguments; they could be found, however, in sociology, literature, and ancient wisdom. If most modern Chinese-language books on this topic give no answers, perhaps we need to go back to the thinkers of an earlier time. Science and technology are reshaping the world, but human nature remains the same. Looking at technological modernization, too much emphasis is given to a multiplicity of opinions, so that is why I have gone back to ancient Chinese thought.

I faced a difficulty with this book that I did not face with my previous two books: I was writing after leaving my position at Alibaba and trying to piece together hours to work in the midst of starting a data technology consulting firm. Just like a computer, the human brain has to be warmed up, too, and I sometimes found that the warmup time was longer than the time I had to write. I often felt guilty about trying to split my time between the book and my business, worrying that I wasn't devoting enough time to either.

This book was only completed with the support of my family and friends. Before anyone, I would like to thank my wife for her understanding and my children for their faith. Secondly, I would like to thank Secretary Chen Gang of Xiong'an New Area, Director Chen Jianhua of Guangzhou Municipal People's Congress, and Deputy Director Mao Guanglie of the Standing Committee of the Zhejiang People's Congress, who have paid close attention to the development of big data technology in China and offered their support to my research and encouragement for my work. I would also like to thank Director Shen Jianrong and Deputy Director He Jun of the Nanjing Development and Reform Commission and Director Xu Wenqing and Director Duan Qingyi of Suzhou Industrial Park. In 2017, I took part in smart city and urban brain projects in both areas, which gave me the opportunity to get real world experience in these highly developed cities. Many cases cited in this book come from my experience and research in Nanjing and Suzhou.

I would also like to say a special thank you to Mr. Tu Xinhui. We met in 2012 because of my work and remained close over the last six years. He has given me indispensable advice on my career development and supported me through unforeseen circumstances in everyday life. He was always ready to offer his heartfelt support. Over the years that I discovered a new direction, new tastes, and new pursuits, I was lucky to have a friend like him.

I would like to thank all of my colleagues, co-workers, and experts in the corporate world. Gao Lutong read the first draft of the book and offered suggestions for revisions and additions. Ge Yumin and Hu Xiaomeng helped me gather materials. A special thank you also to President and Editor-in-Chief Jiang Yongjun and editor Zhang Yingjie of the CITIC Frontier Publishing House of CITIC Publishing Group, who worked meticulously to prepare this book for publication.

Finally, I would like to thank AllBright Law Offices and Dentons for handling all manner of legal affairs, which allowed me to concentrate on my work.

The popularity of mobile phones and WeChat has had a major impact on the reading habits of the public. While I was writing this book, Jiang Yongjun told me to keep my paragraphs brief. It was not worth the trouble and not the best way to write a book. Instead, I hope this book will be a discussion. Even though I am the author and you are the reader, I welcome you to contact me through my WeChat public account. Wherever you are, an Internet connection can bring us together. All you need to do is scan my QR code. Our discussion will also become data. That recording process of our time together will also contribute to our Data Civilization.

Yes, big data and AI will reshape the world, but the warmth of human nature cannot be changed. Things change; other things stay the same.

July 6th, 2018   Tu Zipei

Printed in the United States
by Baker & Taylor Publisher Services